Mathematical Puzzle Tales from Mount Olympus

Mathematical Puzzle Tales from Mount Olympus uses fascinating tales from Greek Mythology as the background for introducing mathematics puzzles to the general public. A background in high school mathematics will be ample preparation for using this book, and it should appeal to anyone who enjoys puzzles and recreational mathematics.

Features

- Combines the arts and science, and emphasizes the fact that mathematics straddles both domains.
- Great resource for students preparing for mathematics competitions, and the trainers of such students.

Andy Liu obtained his doctorate in mathematics as well as a professional diploma in elementary education in 1976, qualifying him officially to teach from kindergarten to graduate school. He was an academic at his alma mater for over thirty years, where he is currently a Professor Emeritus. During that period, he ran a mathematics circle for students in upper elementary or junior high schools. The members had published over fifty papers in scientific journals.

Andy had served as the deputy leader of the USA team and the leader of the Canadian team in the International Mathematical Olympiad. He had chaired its Problem Committee once and was a member three other times. He had authored or co-authored eighteen books in mathematics, and had been credited as an editor or a co-editor for several others.

AK Peters/CRC Recreational Mathematics Series

Series Editors

Robert Fathauer
Snezana Lawrence
Jun Mitani
Colm Mulcahy
Peter Winkler
Carolyn Yackel

Tessellations
Mathematics, Art, and Recreation
Robert Fathauer

Mathematics of Casino Carnival Games
Mark Bollman

Mathematical Puzzles
Peter Winkler

X Marks the Spot
The Lost Inheritance of Mathematics
Richard Garfinkle, David Garfinkle

Luck, Logic, and White Lies
The Mathematics of Games, Second Edition
Jörg Bewersdorff

Mathematics of The Big Four Casino Table Games
Blackjack, Baccarat, Craps, & Roulette
Mark Bollman

Star Origami
The Starrygami™ Galaxy of Modular Origami Stars, Rings and Wreaths
Tung Ken Lam

Mathematical Recreations from the Tournament of the Towns
Andy Liu, Peter Taylor

The Baseball Mysteries
Challenging Puzzles for Logical Detectives
Jerry Butters, Jim Henle

For more information about this series please visit: https://www.routledge.com/AK-PetersCRC-Recreational-Mathematics-Series/book-series/RECMATH?pd=published,forthcoming&pg=2&pp=12&so=pub&view=list

Mathematical Puzzle Tales from Mount Olympus

Andy Liu

CRC Press
Taylor & Francis Group
Boca Raton London New York

CRC Press is an imprint of the
Taylor & Francis Group, an **informa** business

AN A K PETERS BOOK

First edition published 2023

by CRC Press
6000 Broken Sound Parkway NW, Suite 300, Boca Raton, FL 33487-2742

and by CRC Press
4 Park Square, Milton Park, Abingdon, Oxon, OX14 4RN

CRC Press is an imprint of Taylor & Francis Group, LLC

Library of Congress Cataloging-in-Publication Data

Names: Liu, A. C. F. (Andrew Chiang-Fung), author.
Title: Mathematical puzzle tales from Mount Olympus / Andy Liu.
Description: First edition. | Boca Raton : AK Peters/CRC Press, 2023. |
Series: AK Peters/CRC Recreational Mathematics Series | Includes
bibliographical references and index.
Identifiers: LCCN 2022041652 (print) | LCCN 2022041653 (ebook) | ISBN
9781032424545 (hardback) | ISBN 9781032424187 (paperback) | ISBN
9781003362845 (ebook)
Subjects: LCSH: Mathematical recreations. | Puzzles. | LCGFT: Puzzles and
games
Classification: LCC QA95 .L7648 2023 (print) | LCC QA95 (ebook) | DDC
793.74--dc23/eng20221121
LC record available at https://lccn.loc.gov/2022041652
LC ebook record available at https://lccn.loc.gov/2022041653

ISBN: 978-1-032-42454-5 (hbk)
ISBN: 978-1-032-42418-7 (pbk)
ISBN: 978-1-003-36284-5 (ebk)

DOI: 10.1201/ 9781003362845

Typeset in CMR10 font
by KnowledgeWorks Global Ltd.

Publisher's note: This book has been prepared from camera-ready copy provided by the authors.

To

ROBERT GRAVES,

who introduced me to the

Greek Gods and Heroes,

and

NIKOLAY KONSTANTINOV,

who founded the

International Mathematics

Tournament of the Towns

Contents

Contents   ix

Preface

This is a book of recreational mathematics problems set in ancient Greece. It contains 40 tales of gods, mortals and monsters, told by one Prince Eubuleus of Eleusis. The problems do not use mathematical language. They can be understood by anyone with a fair knowledge of high school mathematics.

Since 1992, I have been a vice-president of the International Mathematics Tournament of the Towns. It was founded in 1980 by the great **Nikolay Konstantinov** of Moscow, who recently passed away. The problems are in general hard, but not just for the sake of being hard. Most have elegant ideas behind them. These problems are the kind that one really wants to know how to solve, and not just to get them out of the way. I have chosen 40 problems from this competition and rewritten them as stories about Greek myths and legends.

Eubie, our principal character, was a precocious child, and an active participant in several adventures. He almost fell in when the surface of Earth cracked open. He aced a questionnaire written in Egyptian hieroglyphs without knowing a word of the language. He even became a temporary god! In the three rebellions of the giants, he was employed as a guard, a scout, a messenger and an advisor. In the final chapter, he released Hope from the Pandora Box, giving mortals in despair a reason to live on.

Eubie was first assisted in problem solving by his tutor Silenus the Satyr, a man with two goat legs. Later, Eubie worked with Hebe, the daughter of the Father God Zeus and the wife of the mighty hero Heracles (called Hercules by the Romans but known around Mount Olympus as the *hairy porter*). Silenus and Hebe in turns guided Eubie step-by-step in working out the intriguing problems.

Because the problems were adapted to suit the stories, the original competition problems are listed in the Appendix, along with answers but *no* solutions. The reader may enjoy comparing the problems in the tales with the original problems — I certainly had a lot of fun doing that.

Even for readers primarily interested in preparation for mathematics competitions, I would urge them not to go directly to the Appendix and skip over the puzzle tales. A narrow program of studying past competition problems is not necessarily the best approach. Certainly, the readers would acquire the background knowledge and the basic techniques in an efficient manner, but problem solving calls for much more than that.

Many successful contestants eventually find that they do not seem to be able to do anything other than coaching the next generation of contestants. To break out of this vicious cycle, the readers must be prepared to broaden their interest beyond competition problems. This balances their overall development, and allows much room for free thinking.

Of course, this side interest does not have to be Greek Mythology. However, it is definitely a good choice, with such a wealth of classic literature. The definitive work is *The Greek Myths*, the two-volume treatise by **Robert Graves**, but its derivative, a little book titled *Greek Gods and Heroes*, is my primary reference. I also consulted *The Wonder Book* by **Nathaniel Hawthorne**.

This is not a textbook of Greek mythology. I have taken liberties with the myths. The reader may enjoy comparing the tales here with the stories in *Greek Gods and Heroes*. I had a lot of fun doing that too. What I hope to achieve is to get some readers interested in the subject, and so they will pursue it further on their own.

I also hope that this book will attract the attention of students of the classics who are not interested in mathematics in general. Finding themselves in a familiar environment, they may try to solve some of the problems. If they will go this far, I am convinced that the high quality of the problems will draw them in further.

I am grateful to **George Sicherman** for his meticulous editing, and to **Kate Jones** who has read an earlier draft of the manuscript. Both have made many valuable suggestions. I thank the four anonymous reviewers who have made many constructive criticisms. I thank **Colm Mulcahy** and **Callum Fraser** for encouraging me to submit this book to the A. K. Peters/CRC Recreational Mathematics Series. I thank **Callum Fraser, Mansi Kabra, Robin Lloyd Starkes, Shashi Kumar** and other staff members of **CRC Press/Taylor & Francis** for their administrative and technical support.

Andy Liu,
Edmonton, 2022.

Physical Geography of Ancient Greece

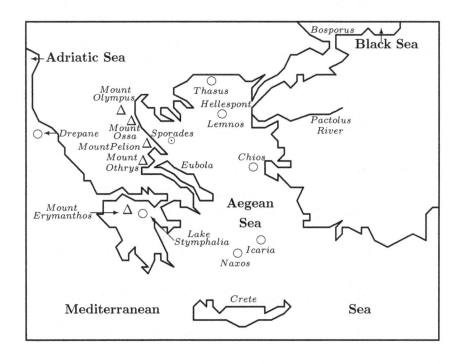

Urban Geography of
Ancient Greece

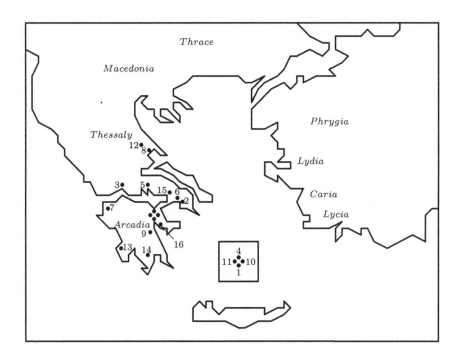

City Index

1	Argos	5	Delphi	9	Lerna	13	Pylus
2	Athens	6	Eleusis	10	Mycenae	14	Sparta
3	Calydon	7	Elis	11	Nemea	15	Thebes
4	Corinth	8	Iolcus	12	Pherae	16	Tiryns

Chapter 1

The Boxing Matches

I, Prince Eubuleus of Eleusis, lived in an exciting time in Greece, when the mortals mingled with the gods, or perhaps it was the other way round. Eleusis was a very poor village. The palace we lived in was just the largest farmhouse in the area. My father the King used a part of it as his office, to run the affairs of his little kingdom. My mother the Queen looked after me and Crown Prince Triptolemus, my elder brother by three years.

When he turned six, Triptie was given the responsibility of looking after the royal cows. A private tutor was hired for his education. Silenus was a satyr, a man with two goat legs. He had a reference letter from Dionysus, the God of Wine and Ecstasy. Dionysus stated that Silenus had been his tutor, and had recently accompanied him and his army on a successful raid of India.

Silenus taught Triptie reading and writing. Triptie's primary interest was mythology, which was a cross between history and mystery. The two of us slept in the same bedroom. At night, he often repeated stories he heard from Silenus during the day.

"Once upon a time," Triptie mimicked Silenus, "there was an Egyptian named Agenor who lived in the land called Palestine. He had a daughter named Europa. One day, she was strolling along the seashore when she was carried away to the north by a snow-white bull. Agenor sent his five sons to go looking for their sister."

Triptie began to tell me the search effort of each of them.

"Did any of them find Europa?" I asked impatiently.

"As it turned out, none of them did."

I lost interest.

When I turned six, I was assigned the duty of feeding the pigs, and allowed to join Triptie to study under Silenus. I learnt to read and write in a very short time. My primary interest was philosophy, particularly logical reasoning. Silenus encouraged me in this pursuit, and gave me many interesting puzzles to solve.

"Eubie," he said, "I believe Triptie told you the story of the search for Europa by her five brothers."

"Yes, sir, but I only remembered that the brothers' names were Pheneus, Cilix, Thasus, Phoenix and Cadmus."

"Well, the second son became a pirate in an area now named Cilicia. The third son became a gold-miner on the island named after him. The fourth son went to North Africa, and built cities. After Agenor's death, he returned to the area in Palestine now called Phoenicia."

"What about the eldest and the youngest sons?" I asked.

"Pheneus spent most of his life in a futile search around the Black Sea, eventually settled down near the Bosporus. Cadmus went to Greece and founded the famous city of Thebes. Here is a little problem for you to solve."

"I am all ears," I said eagerly.

"To raise money for building the city, Cadmus organized a boxing tournament. He put five of his men at each end of a track. At regular intervals, they advanced towards the opposite end. Whenever two of them met, they had a boxing match while all others halted. When the match was over, the two participants turned around and moved in the opposite directions while the other eight resumed moving in the same direction as before. Further meetings resulted in more matches. If a man returned to his starting point, he would retire from the tournament."

"Where did money come in?" I asked.

"Admittance was free, but every spectator had to pay a coin for each match. From this tournament alone, Cadmus raised more than half the money he needed. Can you figure out how many boxing matches took place in that tournament?"

"Let me see," I said. "Let the boxers on the left be labeled 0, 1, 2, 3 and 4, and those on the right be labeled 5, 6, 7, 8 and 9. The first match is between the boxers labeled 4 and 5. Then it gets messy."

I bore down and reconstructed a complete list of all the matches.

```
0 → 1 → 2 → 3 →   ← 4        → ← 6 ← 7 ← 8 ← 9
                     5
0 → 1 → 2 →   ← 3   ← 5        → ← 7 ← 8 ← 9
                 4    → 6    →
0 → 1 →   ← 2   ← 4   ← 6        → ← 8 ← 9
             3    → 5    → 7    →
0 →   ← 1   ← 3   ← 5   ← 7        → ← 9
         2    → 4    → 6    → 8    →
  ← 0   ← 2   ← 4   ← 6   ← 8
     1    → 3    → 5    → 7    → 9    →
0   ← 1   ← 3   ← 5   ← 7
       2    → 4    → 6    → 8    → 9
0   1   ← 2   ← 4   ← 6
         3    → 5    → 7    → 8   9
0   1   2   ← 3   ← 5
             4    → 6    → 7   8   9
0   1   2   3   ← 4        → 6   7   8   9
                   5
```

Each box represented a boxing match. To count them, I draw a slanted 5×5 board on top of my list, as shown in the diagram below. Since there was a box in each square of the board, the number of boxes was $5 \times 5 = 25$.

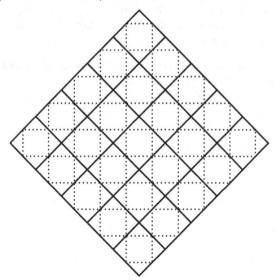

"Your answer is correct, Eubie, but there is a much simpler way of finding it without drawing any diagrams," said Silenus. "Think about it."

Solution

I thought about it for a couple of days. Boxer 0 was in only 1 boxing match, against the returning boxer 1. Boxer 1's first match was against the returning boxer 2. His second match was against the advancing boxer 0. His final one was a rematch with boxer 2. Thus he was in 3 matches altogether.

Analogous reasoning revealed the boxer 2 was in 5 matches, boxer 3 in 7 matches and boxer 4 in 9 matches. By symmetry, the combined number of matches for boxers 5 to 9 was also 1+3+5+7+9=25. Since each match involved two boxers, the total number of matches is 25.

Pleased with myself, I told Silenus what my thoughts were.

"Very good, Eubie, but there is an even simpler solution. You had actually hit upon the key idea when you removed the labels of the boxers from your second diagram. Cast your mind back to the time of the tournament. What did we see after a boxing match? We saw the two participants separating from each other, moving in opposite directions. Since we did not know the identities of the boxers, we would have made exactly the same observation if the two participants had continued in the same directions as before."

"I see now, sir," I said. "The changes in directions, which superficially make the situation more complicated, are actually irrelevant as far as the number of matches is concerned!"

"Indeed," he nodded. "If there were no changes in directions, each of the 5 boxes on the left would meet each of the 5 boxers on the right exactly once. Hence the total number of matches must be $5 \times 5 = 25$."

Chapter 2

The First and Second Labors of Heracles

"Were there any big men born in Thebes, sir?" I asked Silenus the next day.

"Eubie, the Thebans had been asked that question many times before."

"What was the answer then?" I persevered.

"They always said that all they seemed to be able to manage were babies."

After he had recovered from his fit of laughter, Triptie said to Silenus, "You know very well that Eubie is asking whether there are any famous men who were born in Thebes."

"Oh, well," said Silenus, "the most famous Theban has to be Heracles. His mother was Queen Alcmena, the wife of King Amphitryon of Thebes. Zeus, the Father God, fell in love and married Alcmena too. She gave birth to Heracles and his twin brother Iphicles. Angry with her husband's marriages to another mortal woman, Hera, the Mother Goddess, dispatched two monstrous snakes to kill the newborn babies."

"It is hard luck to become famous this way," I said.

"The story is only just beginning. Even as a baby, Heracles was incredibly strong. When the snakes crawled towards the cradle, he caught one in each hand and squeezed them to death. The boys were sent to Mount Pelion to be educated by Cheiron, the king of the centaurs. The centaurs are similar to us satyrs. Cheiron was a man with four horse legs, and extremely wise. Soon, Heracles became an expert in archery, wrestling and boxing."

Triptie said, "I vaguely remember that you have mentioned the *Twelve Labors of Heracles*. What were they?"

5

"Eurystheus, the High King of Greece, wanted to banish Heracles's step-father Amphitryon. Heracles offered to be his slave for 99 months provided that Amphitryon might remain the King of Thebes. Hera saw the chance for her revenge, and advised Eurystheus to set Heracles ten most difficult tasks, to be completed within the 99 month period. She believed that this would finish Heracles off."

"Didn't Triptie mention earlier twelve labors?" I asked.

"I will explain later," said Silenus. "Anyway, the first labor was to kill a lion that had been terrorizing Nemea. Its skin could not be penetrated, but Heracles caught it by the throat and squeezed it to death. He cut the skin off the dead beast and wore it himself thereafter."

"Come on," said Triptie, "give us more juicy details!"

"Let me first talk about the second labor, which was far more dangerous. Heracles was to kill the Hydra in Lerna. This monster had a dog-like body and eight snake heads on long necks. Heracles drew his sword and swung at one of the heads. As soon as it got cut off, two heads grew in its place. After a while, Heracles was facing a 100 headed Hydra."

"He must have got out of it if he was able to do more labors afterwards," I said, "but how?"

"Hera sent a crab to help the Hydra by biting Heracles's foot. She did not realize that she was helping Heracles. Heracles broke the shell of the crab with his club. The crab turned into a young man. He said his name was Iolaus. Some time ago, he had come with three magic swords to kill the Hydra. However, he failed in his mission and was turned into a crab. When Heracles broke the crab's shell, he broke the spell as well, and Iolaus was released."

"How could he help Heracles?" asked Triptie.

"While he was still a crab, Iolaus had thought hard about how to deal with the Hydra. Now he brought along a torch. Every time Heracles cut off one of the Hydra's head, Iolaus would singe the wound to prevent new heads from growing."

"That might have helped if Iolaus had come along in the beginning," said Triptie, "when the Hydra had only eight heads. Now that it had 100 heads, Heracles would not be able to swing his sword fast enough."

"That was where the magic swords came in," said Silenus. "With one swing, the gold sword cut off one-half of the Hydra's current number of heads plus one more, the silver sword cut off one-third of the Hydra's current number of heads plus two more, and the bronze sword cut off one-quarter of the Hydra's current number of heads plus three more."

"What would happen if the current number of heads of the Hydra was not a multiple of 2 or 3?" I asked.

"Good point," said Silenus. "If that happened, Heracles would be turned into a crab as Iolaus had. However, Heracles was careful enough to avoid that. Figure out in how many different ways Heracles could have killed the Hydra, while I satisfy Triptie's thirst for gore by returning to the story of the first labor."

I first considered the last swing. If it was with the gold sword, it cut off one-half of the heads plus one more. Since there were none left, one half of the heads was also one head, so that before the swing, the Hydra was down to two heads. Similarly, if it was with the silver sword, the Hydra was down to three heads, and if it was with the bronze sword, the Hydra was down to four heads. Thus the task could be finished in three ways.

Out of the corner of my eye, I saw Triptie drooling over Silenus's every word. So I decided to put the Hydra out of my mind for the time being, and joined my brother listening to Silenus's graphic description of the killing of the Nemean lion.

Solution

When I came back to the problem of the Hydra. I divided the numbers in 12 classes, 12 being the least common multiple of 2, 3 and 4. If the current number of heads of the Hydra was of the form $12k + 1$, $12k + 5$, $12k + 7$ or $12k + 11$, the mission was impossible. Fortunately, it was 100. I now considered the remaining eight cases.

Case 1. The number was of the form $12k$.
   All three swords might be used. With the bronze sword, the number of heads left would be $9k - 3$, a multiple of 3. With the silver sword, the number of heads left would be $8k - 2$, a multiple of 2. However, with the gold sword, the number of heads left would be $6k - 1$. Heracles must be careful not to use it.

Case 2. The number was of the form $12k + 2$.
   Only the gold sword could be used. The number of heads left would be $6k$, a common multiple of 2 and 3.

Case 3. The number was of the form $12k + 3$.
   Only the silver sword could be used. The number of heads left would be $8k$, a multiple of 4.

Case 4. The number was of the form $12k + 4$.
   Only the silver sword could not be used. With the bronze sword, the number of heads left would be $9k$, a multiple of 3. However, with the

gold sword, the number of heads left would be $6k + 1$. Heracles must be careful not to use it.

Case 5. The number was of the form $12k + 6$.
Only the bronze sword could not be used. With the silver sword, the number of heads left would be $8k + 2$, a multiple of 2. With the gold sword, the number of heads left would be $6k + 2$, also a multiple of 2.

Case 6. The number was of the form $12k + 8$.
Only the silver sword could not be used. With the bronze sword, the number of heads left would be $9k + 3$, a multiple of 3. With the gold sword, the number of heads left would be $6k + 3$, also a multiple of 3.

Case 7. The number was of the form $12 + 9$.
Only the silver sword could be used. The number of heads left would be $8k + 4$, a multiple of 4.

Case 8. The number was of the form $12k + 10$.
Only the gold sword could be used. The number of heads left would be $6k + 4$, a multiple of 2.

Starting with 100 heads, Heracles must use the bronze sword to leave 72 heads. Now he had two choices.

Case 1. He used the silver sword.
This left 46 heads. His further actions were forced, using the gold sword to reduce the number of heads to 22, 10 and 4 in succession, and finishing the job with the bronze sword. This yielded 1 way.

Case 2. He used the bronze sword.
This left 51 heads. So he must use the silver sword to leave 32 heads. Once again, he had two choices.

   Subcase 2(a). He used the gold sword.
   This left 15 heads. So he must use the silver sword to leave 8 heads. This might be reduced to 3 by using either the bronze sword or the gold sword. The job would then be finished with the silver sword. This yielded 2 ways.

   Subcase 2(b). He used the bronze sword.
   This left 21 heads. So he must use the silver sword to leave 12 heads. This might be reduced to 6 by using either the bronze sword or the silver sword. This might be reduced further to 2 by using either the silver sword or the gold sword. The job would then be finished with the gold sword. This yielded 4 ways.

In summary, there were 7 different ways for Heracles to kill the Hydra.

Chapter 3

The Third and Fourth
Labors of Heracles

"The third labor," Silenus continued the next day, "was for Heracles to capture a hind which had come over from Ceryne in North Africa. It was both graceful and intelligent. Heracles chased it up hill and down dale all over Greece for over a year."

"That was a large chunk of his time allowance of 99 months," my brother pointed out.

"Good point, Triptie. Heracles finally realized that he must use a different approach. He marked 18 evenly spaced spots in a row. He knew that the hind loved apples, and put one on each spot. The hind would not go away until it had eaten all of them. Starting from one of the spots, it would eat the apple there. Then it jumped in either direction, with spans of 8, 9 or 10 steps. In other words, there must be 7, 8 or 9 other spots in between. It would only jump to spots which still had apples on them. Heracles hoped to seize it while it was figuring out how to get all the apples."

"You said that the hind was intelligent, sir," I said. "If it figured things out quickly, Heracles would not have time to carry out his plan."

"Indeed, that was what happened. So in each of the following days, Heracles added one more spot with an apple on it. After a few days, the hind was caught. Heracles brought it to Eurystheus, who then set the hind free."

"There is not a single drop of blood in the whole story," Triptie complained. "I hope the story about the fourth labor will be more to my taste."

"It will not be to your brother's taste. Eubie, why don't you go off and think about the hind and the apples?"

I wrote down the following chart, showing how the hind might jump.

Spans of Jumps	Spots								
	1	2	3	4	5	6	7	8	9
8 steps	9	10	11	12	13	14	15	16	17,1
9 steps	10	11	12	13	14	15	16	17	18
10 steps	11	12	13	14	15	16	17	18	–

Spans of Jumps	Spots								
	10	11	12	13	14	15	16	17	18
8 steps	18,2	3	4	5	6	7	8	9	10
9 steps	1	2	3	4	5	6	7	8	9
10 steps	–	1	2	3	4	5	6	7	8

The hind would get all the apples if it started from spot 9 or 10 and followed the sequence of jumps: 10–1–11–2–12–3–13–4–14–5–15–6–16–7–17–8–18–9.

Then I wrote down a similar chart for the case with nineteen apples.

Spans of Jumps	Spots									
	1	2	3	4	5	6	7	8	9	10
8 steps	9	10	11	12	13	14	15	16	17,1	18,2
9 steps	10	11	12	13	14	15	16	17	18	19,1
10 steps	11	12	13	14	15	16	17	18	19	–

Spans of Jumps	Spots									
	11	12	13	14	15	16	17	18	19	
8 steps	19,3	4	5	6	7	8	9	10	11	
9 steps	2	3	4	5	6	7	8	9	10	
10 steps	1	2	3	4	5	6	7	8	9	

This time, the hind could start from any spot since the workable sequence closed into a loop: 10–1–11–2–12–3–13–4–14–5–15–6–16–7–17–8–18–9–19–10.

It seemed to me at this point that the hind could always get all the apples. I just had to find a sequence of jumps for it to follow. However Silenus had said that the hind was captured after a few days. I had no idea how to handle that situation. I did not suppose I could exhaust all possible sequences of jumps. I was getting tired, and decided to think no more about it for the day.

At night, Triptie told me the gist of the story of the fourth labor of Heracles.

"Once upon a time, there was a wild boar which lived in Mount Erymanthos. It was a huge beast with an elephant's tusks and a rhinoceros's hide. Heracles chased it up and down the mountain in winter, until it got stuck in a deep snowdrift. Then he jumped on it and tied it up. He brought it to Eurystheus, who was scared to death by the sight of the creature."

Solution

The next day, I continued my attack on the problem for the case with 20 apples. The situation was very similar to the case with 18 apples. This time, the hind must start from spot 1 or spot 20. The workable sequence was: 1–11–2–12–3–13–4–14–5–15–6–16–7–17–8–18–9–19–10–20.

For the case with 21 apples, I modeled on the case with nineteen apples: 11–1–12–2–13–3–14–4–15–5–16–6–17–7–18–8–19–9–20–10–21–11.

Suddenly, I realized that this was not a workable loop, because ten of the jumps had spans of 11 steps!

I was both excited and apprehensive. This might be the case in which the hind was captured. Not having found a workable sequence myself did not mean that the hind could not either. I had to find a proof that it indeed could not.

I thought and thought, and came up with many different ideas. Unfortunately, none of them bore fruit. I went to Silenus for help.

"Eubie, you may introduce some coloring of the spots," he suggested.

"The most natural coloring scheme, sir," I said, "is to paint the spots alternately white and black. There will be 11 white spots and 10 black ones."

$$2 \quad 4 \quad 6 \quad 8 \quad 10 \quad 12 \quad 14 \quad 16 \quad 18 \quad 20$$
○ ● ○ ● ○ ● ○ ● ○ ● ○ ● ○ ● ○ ● ○ ● ○ ● ○
$$1 \quad 3 \quad 5 \quad 7 \quad 9 \quad 11 \quad 13 \quad 15 \quad 17 \quad 19 \quad 21$$

"What are the jumps between spots of the same color?"

"The jumps with spans of 8 or 10 steps are between spots of the same color, while the jumps with span 9 are between spots of different colors. However,", I said after a while, "this is not leading me anywhere."

"Why not try varying the coloring scheme?"

It was time for me to feed the pigs. I drew the diagram below on the sand.

$$2 \quad 4 \quad 6 \quad 8 \quad 10 \quad 12 \quad 14 \quad 16 \quad 18 \quad 20$$
○ ○ ○ ● ● ● ○ ○ ○ ● ● ● ○ ○ ○ ● ● ● ○ ○ ○
$$1 \quad 3 \quad 5 \quad 7 \quad 9 \quad 11 \quad 13 \quad 15 \quad 17 \quad 19 \quad 21$$

I found all the jumps between two white spots. They all had spans of 8 or 10. The complete list was: 1–9, 7–15, 13–21, 3–13 and 9–19. The jumps between two black spots were: 4–12, 10–18 and 6–16.

Suppose the hind made a jump between two spots of the same color. I would merge those two spots into one. Suppose the hind followed the sequence 1–9–19 or 3–13–21. I would merge all three spots into one. Then the merged spots must be alternating in color.

At the start, there were 12 white spots and 9 black spots. Without any merger, these two kinds of spots could not be alternating. However, with mergers, they could. So another approach came to a dead end.

I reported my failure to Silenus later in the day.

"Don't give up, Eubie. There are other coloring schemes.'

Using the fact that $21 = 3 \times 7$, I introduced a third coloring scheme.

$$\begin{array}{cccccccccc} 2 & 4 & 6 & 8 & 10 & 12 & 14 & 16 & 18 & 20 \\ \circ \;\; \circ & \circ & \circ & \circ & \circ & \bullet \;\; \bullet & \bullet \;\; \bullet & \bullet \;\; \bullet & \circ \;\; \circ & \circ \;\; \circ \;\; \circ \;\; \circ \\ 1 & 3 & 5 & 7 & 9 & 11 & 13 & 15 & 17 & 19 & 21 \end{array}$$

This time, there were 14 white spots and 7 black ones. Since it was impossible for the hind to jump between two black spots, the number of merged black spots remained 7.

The only possible jumps between two white spots were: 5–15, 6–15, 6–16, 7–15, 7–16 and 7–17. Hence none of the white spots 1, 2, 3, 4, 18, 19, 20 and 21 was involved in any merger. Even if all the white spots 5, 6, 7, 15, 16 and 17 were merged into a single one, the number of merged white spots was still 9. Hence the merged white spots and the merged black spots could not be alternating in color.

I reported my success to Silenus. He was delighted.

"Well done, Eubie," he said. "Indeed, the Ceryneian hind was captured when Heracles laid out 21 apples. It might not have realized that the task was impossible. In the futile search for a non-existent solution, it was surprised by Heracles."

Chapter 4

The Fifth and Sixth Labors of Heracles

The fifth labor of Heracles was very different in nature from any of the first four. He was told to clean out the filthy 10×10 cattle yard of King Augeias of Elis in a single day. It had not been tended to in over thirty years. Even I found the story somewhat tame, not to mention smelly. Moreover, it was not a task suited to the talents of Heracles.

Heracles came up with a brilliant idea. He planned to divert the rivers nearby to wash the cattle manure away from the field. However, he did not want to flood the whole area all at once in case things got out of control. So he built a wall around the yard and around the northwestern quarter of the yard. Finally, he built some dykes along the sides of the squares in the other three quarters, as shown in the diagram below.

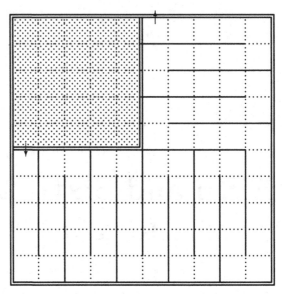

Silenus explained, "Building the walls and the dykes was easy work for Heracles, and he completed that by late afternoon. He let water into the northwest quarter to clean out that part, and enough to cover the rest of the field. Then he broke the inner wall at the southwest corner of the northwest quarter."

"That was clever," I said in admiration. "Now the water just wound around the dykes and cleaned out the rest of the field."

"Heracles did not reckon that the 75 steps the water needed to go through the channel would take so much time. Fortunately, the whole field was flooded by the end of the day. He released the dirty water by breaking the outer wall near the middle of the north edge."

"Let's move onto the sixth labor then," said Triptie impatiently.

"Heracles was told to drive off the man-eating birds from Lake Stymphalia. They looked like cranes, but with brass feathers. The lake was marshy, so that Heracles could not approach the birds. He shot at them, but his arrows simply bounced off their feathers."

"What did he do then, sir?" I asked.

"He had some unexpected help, Eubie," said Silenus. "It came from my father Pan, a minor god. He happened to be hiding in a cave near the lake. When Heracles passed by and startled him, he let out a terrifying yell."

"How did that help Heracles?" asked Triptie.

"The cry sent the birds into the air, mad with what was known as a *panic* fear. They had no brass feathers on their undersides. Many of them were shot dead by Heracles, and the rest flew away and never returned."

Later that day, my thoughts returned to the cleaning of the cattle field. I suddenly realized that if Heracles flooded the field in two separate channels, as shown in the diagram below, the time for the water to wind around the dykes could be cut down to 39 steps, the length of the longer channel.

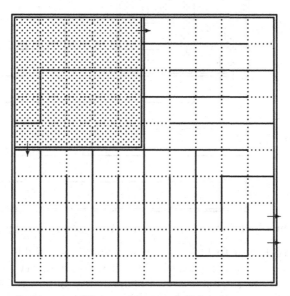

I reported my findings to Silenus. He was delighted.

"Eubie," he nodded with approval, "you are not just improving in problem solving fast. You have begun to propose problems as well. Can you find further improvements on the amount of time saved?"

Solution

"Let me see," I said. "Consider the northwest corner square. Water there takes five steps to get out of the northwest quarter. So it will take another 15 steps to get out of the field. Hence the least number of steps to drain the whole field cannot be less than 15."

"Is it possible to do that?" Silenus asked.

I went away to do more thinking. Finally, I came up with a workable scheme. I showed Silenus my diagram.

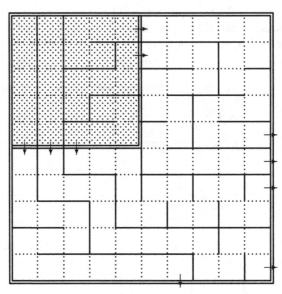

I thought I had settled all the issues, but Silenus found a new twist.

"Suppose I wish to have as many channels as possible. Of course they need not all have the same length. Also, it is not necessary to drain the water through breaks in the outer wall. I allow underground cavities which permit draining within the field. Eubie, what is the maximum number of channels we can have?"

Seeing that I could not solve the problem right away, Silenus told me to sleep on it.

"Tell me a bedtime story then, sir."

"We satyrs are not handsome creatures. My father, the God of the Shepherds and Hunters, is ugly even by our standard. That is why he spends most of his time hiding in caves. I heard that before he married my mother, he fell in love with a nymph named Syrinx. When he tried to kiss her, she *panicked* and turned herself into a reed forever."

"I know of a musical instrument called the *Pan pipes*. It is made from reeds. Does that have anything to do with your father?"

"Yes, he is credited with its invention. He dedicates it to the memory of Syrinx."

I woke up in the middle of the night and solved the new problem Silenus proposed. Since only 9 squares in the northwest quarter were adjacent to the inner wall, the number of channels could not be greater than 9. I came up with the scheme shown in the diagram below, the circles representing underground drainage. The whole process still took only 15 steps.

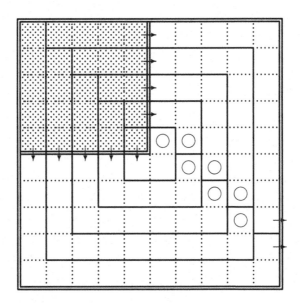

Chapter 5

The Seventh and Eighth Labors of Heracles

"Let us return to the story of Europa," Silenus said to us.

"No," said Triptie. "I want more stories about the labors of Heracles."

"Actually, what I am about to tell you is related to the seventh labor. Eubie, you remember that none of Europa's brothers ever found her."

"Yes, sir. She seemed to have evaporated from earth."

"The snow-white bull which carried her off was none other than Zeus in disguise. He headed north for a while before turning around, leaving behind a false trail which fooled Pheneus badly. Cadmus had a better sense of what had transpired, but he did not go far enough south. Zeus took Europa to Crete and married her there."

"What else would you expect from Zeus?" Triptie said. "What has this got to do with the seventh labor?"

"Be patient," said Silenus. "Europa gave birth to a son Minos, lived to a ripe old age and gave her name to our continent, now called Europe. When Minos grew up, he wanted to become the King of Crete. He told the Cretans that the gods would send a sign to support his claim. Sure enough, a snow-white bull came ashore from the sea."

"Was it still Zeus in disguise?" I asked.

"No, but Minos got his wish. After his coronation, he should have sacrificed the bull to Zeus. However, he liked the beautiful animal and sacrificed a different bull. Zeus punished him by letting the bull run off and cause havoc all over Crete. As the seventh labor, Eurystheus sent Heracles to capture it."

"Why would that count as a labor? Heracles wouldn't have too much trouble with that, would he?"

"If he were to kill the bull, it would indeed have been a simple task. However, he was obliged to take the bull safely across the sea to Eurystheus."

"How did he manage it?" I asked.

"To escape from Heracles, the bull hid in a wood. There were three clearings, and some low bushes in each. I have drawn a diagram for you."

"I take it that the small black dots represent the bushes. What about the large black dot, the large targeted dot and the lines?" Triptie asked.

"At dusk, the bull came into an opening to get some exercises. He always appeared at the bush represented by the large targeted dot. The large black dot represented a bush which afforded Heracles the most concealment, and he always started there. Heracles moved first, the bull responded, and they took alternate turns thereafter. In his turn, Heracles could move from one bush to another if the dots representing them were joined by a line, with no other dots in between. The same applied to the bull in its turn."

"I don't think Heracles could catch the bull in the first clearing," said Triptie after a while. "Let me redraw the diagram as a 4 × 4 board, with the squares painted black and white so that no two squares sharing a common side have the same color."

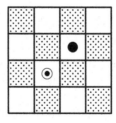

"How would that help?" I asked.

"If Heracles could land on the bull, they must be in the same square, and hence in squares of the same color. This was the case at the start. After each turn of Heracles, wherever he might move, they would be in squares of different colors. It follows that Heracles would not succeed. Moreover, the bull could then move to a square of the same color as the one occupied by Heracles, as long as it was not that very square."

"Excellent," said Silenus. "As it turned out, Heracles would fail in his task in one of the remaining two clearings. Eubie, can you see which one it would be?"

"I think so," I said. "It would be the third one. At first I try to copy the clever idea of my brother. However, with the corner bushes removed, his coloring argument falls apart. Then I have my own idea. I observe that the first clearing is symmetric about its center, and the starting positions of Heracles and the bull are also symmetric. Wherever Heracles moved, the bull could move to the symmetric bush. Removing opposite corner bushes does not invalidate this argument, so that it applies to the third clearing as well as to the first."

"Great. I can tell you that Heracles did succeed catching the bull in the second clearing. Can either of you see how?"

We were both stumped for the time being.

"Let us continue with the eighth labor," Triptie suggested.

"This time, Eurystheus sent Heracles to capture the four man-eating mares of King Diomedes of Thrace. Unfortunate strangers who wandered into his kingdom would be taken by him and fed to his mares at night."

"I am going to have *nightmares*," I gulped.

"Heracles sailed to Thrace and landed near the stables of Diomedes. He overpowered the mares and harnessed them. He drove them to the seashore and attempted to take them on board. However, his ship was too small for all four of them. Meanwhile, an alarm had been raised, and Diomedes came after Heracles at the head of his palace guards. Heracles left the mares in charge of a boy named Abderus, and met the enemy head on."

"Was that wise?" I questioned.

"No," said Silenus. "Abderus was unable to control the mares. It was a short fight, and Diomedes was stunned when Heracles hit him on the head with his club. Unfortunately, by that time, Abderus had been eaten by the mares. Outraged, Heracles fed Diomedes to them."

"*Served* him right!" said Triptie.

"Heracles then instituted an annual funeral games in honor of Abderus, before driving the mares over land through Macedonia back to Eurystheus."

The story was much to Triptie's liking, and he slept soundly on it. Unable to do so, I sat up and worked on how Heracles could have caught the Cretan bull in the second clearing.

Solution

Neither the coloring nor the symmetry ideas seemed to work here, but I could think of nothing else. Perhaps it would work if I somehow combined the two. I first redraw the diagram of the second clearing. I omitted the northeast corner square from Triptie's diagram and linked its two former neighbors by a quarter circle, to indicate that direct movement between them was permitted. This new move must be the reason for the success of Heracles.

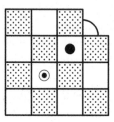

To make use of it, Heracles should move north, follow the quarter circle and then move west back to where he was. These three moves could also be taken in the reverse order.

What had he achieved? The bull could not maintain the symmetry about the center of the clearing. However, that did not mean that it would be captured, as it might have other strategies. However, after the third turn of each party, they would be in squares of opposite color, which spelt doom for the bull. The only way it could restore the original situation was to move along the quarter circle itself, but Heracles was in a position to prevent it from getting there.

Chapter 6

The Ninth and Tenth Labors of Heracles

The ninth labor was another strange one. Heracles was to get a golden belt from Queen Hippolyte of the Amazons, and bring it back as a present for Eurystheus's daughter.

"Yuck!" Triptie almost threw up.

I agreed with him to some extent. I asked Silenus, "It is a most unusual job for a man, sir. Is there any danger?"

"For ordinary people, Eubie," said Silenus, "but not for Heracles. The Amazons are a tribe of women warriors who live on the southern coast of the Black Sea, next to a tribe of men called the Gargareans. Once a year, the Amazons mate with their neighbors in order to perpetuate both races. All boys born are sent back to their fathers. When the girls grow up, their right breasts are cut off because they get in the way when they try to shoot arrows with bows."

"They must be very interested in fighting, then," remarked Triptie, his interest somewhat revived.

"Queen Hippolyte fell in love with Heracles, and was about to give him the belt. However, Hera spread a rumour among the Amazons that Heracles had come to kidnap their queen. The angry Amazons came to rescue her. The resulting melee was most confusing. Though Heracles pervailed, Hippolyte was accidently killed by an arrow shot by one of her people. Sadly, Heracles took the belt from her dead body, and brought it back to Eurystheus."

"Get on with the tenth labor," urged Triptie, disappointed once again.

"This time, Heracles was asked to steal a herd of red cows from King Geryon who lived on an island in the Western Ocean beyond the Mediterranean Sea. Geryon had three bodies joined at the waist, but only one pair of legs. When Heracles sailed to the western end of the Mediterranean Sea, he found that his voyage was blocked by an isthmus joining Africa to Spain. It was of length 20 and width 1."

"Did he have to haul the boat over land before he resumed sailing again?" I asked.

"That would be too much trouble," said Triptie. "I would cut through the isthmus."

"That was precisely what Heracles did. He got onto the isthmus and cut it up into 20 blocks with square bases. In order for the boat to pass through, the 18 blocks in the middle had to be out of the way. He lifted one of the blocks and threw it on top of an adjacent one, making a pile of height 2. This took him the better part of a day."

"I would say," I said.

"In each subsequent day, he lifted an entire pile, whatever the height might be, and threw it on top of another pile at a distance equal to the height of the pile being thrown. Eventually, there were two piles of height 10, one at each end of a new strait of width 18. These two piles are now called the Pillars of Heracles."

"That took 18 days," I said, "and it must have drained a lot of strength from Heracles."

"Yes, but he had a restful voyage before he got to the island where Geryon lived. He was first attacked by Orthrus, a fierce two-headed dog, but Heracles shot it dead with a single arrow, piercing both heads. Then Geryon rushed like a row of three men. Heracles waited until Geryon's three heads were lined up properly, and killed him by shooting an arrow through all of them."

"Hurrah for Heracles!" Triptie cheered.

"Heracles drove the red cows across the Pyrenees and along the south coast of France," Silenus continued. "At the Alps, however, he met Hera in disguise, and she misdirected him to turn right too soon. He went quite a way down before realizing that he was not in Greece but Italy. He turned back and delivered the cows to Eurystheus just as the 99 months expired."

"That was a close call," I said. "Heracles should now be set free. Why are there two more labors, and what are they?"

"First, Eurystheus said that in the second labor, Heracles had help from Iolaus."

"That was not fair," Triptie said. "Heracles did not call on Iolaus. Hera sent him as a crab to bite Heracles, but her evil plan backfired."

"Heracles made that point too, but Eurystheus brushed his objection aside. Eurystheus also complained that the fifth labor was really done by the river gods."

"How did Eurystheus expect Heracles to clean the field without water?" I said.

"Well, Eurystheus was the High King, and his words were final. So two replacement labors were set, with extra time allowed. However, we will leave these stories to another day."

Later that day, I suddenly realized that I had overlooked the problem of how the Pillars of Heracles were formed. It seemed reasonable to assume that the 10 blocks to the north formed the pillar in Spain, while the 10 blocks to the south formed the pillar in Africa. Thus I needed handle only the northern half or only the southern half. In any case, both halves could be handled in identical fashion.

My first approach was to make the first available move from left to right in the diagram below.

	2	1	1	1	1	1	1	1	1	Shore
		1	3	1	1	1	1	1	1	Shore
			4	1	1	1	1	1	1	Shore
				1	1	1	5	1	1	Shore
					2	1	5	1	1	Shore
						1	7	1	1	Shore
							8	1	1	Shore
							8		2	Shore

Had the width of the isthmus been 16, the problem would have been solved. Now the only thing I could do was to throw the pile of height 2 on top of the pile of height 8.

							10			Shore

I did get a pillar of height 10, but it was away from the shore.

Solution

I convinced myself that the problem was not really all that hard, and resisted the urge to seek out Silenus. After working on it for another while, I hit upon the right idea, going back and forth. I started by throwing the fourth block on top of the fifth one, forming a pile of height 2. Thereafter, I always threw the one of height greater than 1 in the direction opposite to the preceding move. Thus I built up piles of increasing height, first to the right, and then to the left, alternating until I got a pillar of height 10 by the shore.

1	1	1	1		2	1	1	1	1	Shore
1	1	1	3			1	1	1	1	Shore
1	1	1				4	1	1	1	Shore
1	1	5					1	1	1	Shore
1	1						6	1	1	Shore
1	7							1	1	Shore
1								8	1	Shore
9									1	Shore
									10	Shore

Chapter 7

The Loss of the Pigs

The true beginning of my adventures occurred when I was nine years old. I took the pigs to a 9×9 field not far from home. I let them run around, looking for food. There was a stone in the top right corner square. I sat on it and drew diagrams in the sand with a tree branch, trying to figure out Silenus's latest puzzle.

Suddenly, I heard some rumbling noise above my head. I looked up and saw a chariot racing across the sky. This was nothing remarkable as it was always full of all kinds of flying objects. So I paid no more attention to it.

The next moment, an earthquake came out of nowhere. Some of the 81 squares of the field cracked up along both diagonals. Along with the edges of the field, these cracks isolated a few pieces of land from the square with the stone. They fell beneath the earth, and this led to the collapse of the whole field except for the square with the stone.

All the pigs fell in right before my eyes. Weak-kneed, I hobbled home in a blind fear. Seeing the state I was in, my mother put me straight to bed. In total exhaustion, I fell into a deep slumber.

"Eubie," she asked me the next day, "what happened yesterday, and where are the pigs?"

"The pigs are gone, mother," I began to cry. "It is not my fault. The earth just opened up and swallowed them. I was scared to death."

"Don't cry, Eubie," Triptie comforted me. "At least you have not been swallowed."

"I almost was! There was a chariot in the sky. The people in it must have made the earth open up, because the chariot went through the opening and disappeared."

"Who were in the chariot?" asked my mother.

"Let me think," I said. "I remember a big man with a frightening dark face, and a young woman with a frightened pale face. The earth closed up again after the chariot went through."

"Going to Tartarus the Underworld in broad daylight," mused Triptie. "That man must be Hades, the God of Death. Eubie, you had a close call."

With the pigs gone, I could afford to rest for a few days. Silenus had told me to take a break with my studies. He came to visit me after he had given Triptie his lessons.

"Are you feeling better now, Eubie?"

"I am still a bit shaky, sir."

"Let me get the picture of what happened. You said that some of the squares of the field cracked up along both diagonals. If at least one piece of land is isolated from the square with the stone, then the whole field except that square will collapse. I have drawn a hypothetical situation and shaded the isolated pieces of land. Am I correct?"

He showed me the diagram below.

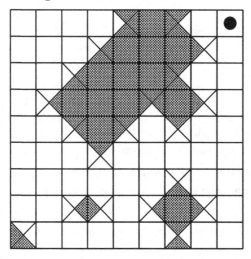

"Yes," I said. "From the small examples in your diagram, I see that if a square at a corner or along an edge cracks up, at least one piece of land will be isolated from the square with the stone. This is also true if two squares sharing a common side both crack up."

"What is the maximum number of squares that may crack without isolating any piece of land from the square with the stone?"

"The squares at the corners or along the edges of the field must remain intact. Hence the maximum is at most 49. However, some of the squares of the inner 7×7 field must also remain intact. I don't see right away how many that need to be."

"Try to crack up as many squares as you can without collapsing the whole field. When you feel that you have found the optimal way, try to prove that it is indeed optimal."

Solution

I had observed that for any two squares sharing a common side, at least one must remain intact. I painted the squares of the field black and white, so that no two squares of the same color shared a common side.

For the optimal result, I crack up only squares of the same color. For the next few days, I kept drawing. The best was the diagram below, with 21 cracked squares. The shaded squares were the black ones. They were not isolated pieces of land.

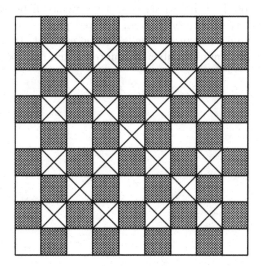

I showed it to Silenus.

"Is 21 the maximum, Eubie?" Silenus asked.

"I believe so, sir, and here is my reasoning," I explained. "I divide the inner 7×7 field into four 3×4 rectangles plus a single square at the center, as shown in the diagram below.

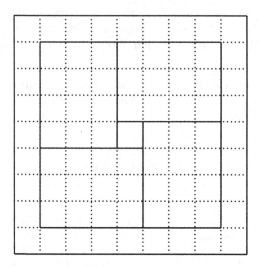

"Why do you do that?"

"Consider a 3 × 4 rectangle. In each of the 3 rows of length 4, we can crack at most 2 squares. However, if we crack exactly 2 squares from each row, with no 2 of the 6 adjacent, then they must all be of the same color. However, four of them will enclose an isolated piece of land. It follows that in any 3 × 4 rectangle, at most 5 cells can be cracked. It follows that in the 9 × 9 field, the maximum number of squares which may be cracked is indeed 4 × 5 + 1 = 21."

Chapter 8

The Planting of Barley

Nine days passed. I had recovered sufficiently to rejoin Triptie in our studies. A wandering old woman came by the palace. She was weak and frail. She said that she had not eaten for days. Feeling sorry for her, my mother gave her some food and barley-water, and offered to let her stay with us until she had regained some strength. She accepted both gratefully, and in return, agreed to help my mother look after me.

The old woman was good company, and gradually, I told her my frightening experience. Suddenly, her face lit up, and she looked a totally different person.

"Eubie," she said to me, "I am Demeter, the Goddess of all Useful Fruits, Grasses and Grains. I recently lost my daughter Persephone. That was why I disguised myself as an old woman, looking everywhere for her. I am grateful to you for this important piece of information. I am sure that the young woman you saw in the chariot is Persephone."

"My brother said that the big man must be Hades, madam," I added, remembering Triptie's remark.

"I am certain of that. He is actually my elder brother, but he had eyes on Persephone for some time. Your family has been very kind to me, and I will return. I must now go off to look for concrete evidence, to bring before the Father God Zeus, another elder brother of mine."

My father and mother came to say good-bye to her, and apologized for not having accorded her the honor she was due. She brushed that aside, and gave them some advice.

"I notice that all your fields are 9×9. Plant barley in as many squares in each field as you can. However, if you plant barley in the square marked with

31

a circle in the diagram below, you must not do so in any of the 12 squares around it."

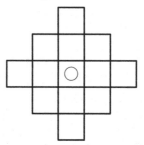

After she had gone, my parents asked me to consult Silenus as to how we should follow Demeter's advice. We worked on it for a little while.

"Eubie, a good way for solving problems is called the *Divide and Conquer* method," he said. "Let us divide the 9×9 field into nine 3×3 subfields. Can you see in at most how many squares of each subfield you can plant barley?"

"Just 2, sir," I answered at once. "Even if you plant barley in a corner square, that eliminates 5 other squares of the subfield. In the remaining 3, we can plant barley in at most 1 of them. It may be the square at the opposite corner, or a square sharing a common side with that square."

"This means that the maximum is at most 18. Can it actually be 18?"

We tried very hard finding a configuration of 18 squares that would satisfy Demeter's condition. The best we could come up with was a configuration of 17 squares, as shown in the diagram below. Since he could not afford to wait, my father went ahead with the planting.

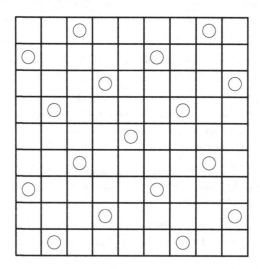

About a month after Demeter's departure, Triptie began to notice that there was not enough grass for the royal cows to eat. Then the trees stopped bearing fruit, and everything stopped growing except for barley. It was a difficult time for all because barley was planted in only a few places in Greece. However, because of Demeter's advice, we had plenty in Eleusis. As a result, we managed to do reasonably well.

My father, mother and Triptie were happy, but Silenus and I were not. We were still wondering if the maximum could actually be 18.

Solution

After quite a while, Silenus began to suspect that 18 was not attainable, and we turned our attention to looking for a proof. We assumed that we could plant barley in 18 squares, and looked for a contradiction.

Silenus divided the argument into two cases, in that we would either plant barley in the middle square of at least one of the four edges of the 9×9 field, or we would not do so in the middle square of any edge.

I took on the first case, and assumed by symmetry that the middle square of the north edge was planted with barley. Since we must plant barley in 2 squares of each subfield, the other squares in the north subfield must be either the southeast corner square or the southwest corner square. Again by symmetry, I could assume that it was the southwest corner square, as shown in the diagram below.

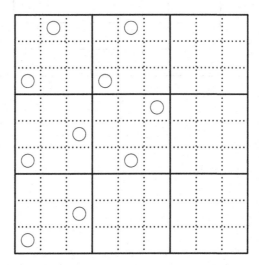

This forced the choices of the 2 squares in the northwest subfield. This placement, along with that in the north subfield, forced the choices of the 2

squares in the west subfield. In turns, the placements in the central subfield and the southwest subfield were also forced. Now it was not possible to choose 2 squares in the south subfield.

Elated, I ran to Silenus and reported my success.

"Excellent work, Eubie," Silenus nodded with approval. "Meanwhile, I have also been successful with the second case. Assuming that none of the central squares of the four edges was planted with barley, I divide the remaining part of the field into 17 crosses, some missing a square."

"Let me see, sir," I said. "We cannot plant barley in more than 1 square in any cross. So the maximum is at most 17."

"In fact, 17 is the maximum since we have earlier found a configuration of 17 squares that satisfy Demeter's condition. Actually, those 17 squares are the central square of the crosses in the diagram below."

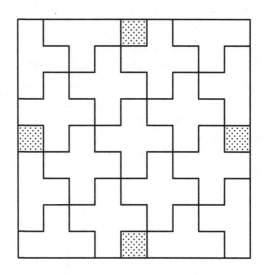

Chapter 9

The Fruit Pies

Having met Demeter, Triptie and I asked Silenus to tell us more about the gods and the goddesses.

"Once upon a time," he began, "the world was ruled by the titans, under King Cronus and Queen Rhea. They have five offspring. The eldest is a girl Hestia, the Goddess of the Home. The next three are Hades, Poseidon and Zeus. Demeter is the youngest. Then Zeus led a palace revolt. The titans were deposed and exiled. The new rulers built a huge palace on Mount Olympus, the highest mountain in Greece. They call themselves the Olympians. To run the affairs of the world, they form an Olympian Council with 12 seats."

"Sir," I asked, "why is Zeus, the youngest of the three brothers, the Father God?"

"He started the revolt, Eubie," said Silenus. "Cronus had imprisoned the one-eyed cyclopes in Tartarus. Zeus sent Hades to release them. In gratitude, the cyclopes gave Hades a helmet of invisibility and Poseidon a trident. They made thunderbolts for Zeus, a special weapon which could deliver *prehumous* cremation."

"My god!" exclaimed Triptie ambiguously.

"He is said to be a strong, brave, stupid, noisy, violent and conceited god. He is always careful in making sure that his family would not be able to get rid of him, as he had got rid of his father."

"It seems that he is not as stupid as you think," I said.

"The three brothers decided to share the earth among them," continued Silenus. "Poseidon married Amphitrite, the Goddess of the Seas and the Rivers, and then took over all her titles. Zeus banned Hades from visiting Olympus."

35

"Heavens forbid!" remarked Triptie. "How could Hades have designs on his own niece?"

"Morals are for the mortals," Silenus laughed. "Incest among the gods happens all the time, intergenerational or otherwise. It is widely believed that Zeus is the father of Persephone."

Half a year went by. Then Demeter returned to our palace.

"I have come here to wait for my daughter," Demeter told us. "When Hades asked for permission to marry Persephone, Zeus did not want to offend him, but he also did not want to offend me. So he just winked. Hades took it as tacit approval, and carried out his evil plan. I had gathered enough evidence and witnesses, but Zeus still prevaricated."

"What did you do then?" my father asked.

"To force his hand, I stopped everything from growing. Finally, he decided to make Hades return Persephone to me, as long as she had not tasted the Food of the Dead. Hades is bringing her back here in his chariot."

"I must keep the royal cows out of harm's way then," said Triptie. "The earth is going to open up again."

Sure enough, the big man and the young woman I saw before came back in the same chariot the next morning. There was a tearful reunion between mother and daughter. To celebrate the occasion, my mother baked 15 fruit pies and arranged them along the rim of a round plate.

Then Hades grinned and said, "By the way, although Persephone never ate any meal while in Tartarus, my gardener had witnessed her eating 7 pomegranates from my underground orchard. She has to marry me and go back to Tartarus."

"The deals is off!" screamed Demeter. "Nothing grows again!"

During the impasse, my mother took me aside and said, "Eubie, one of the 15 pies is a pomegranate pie, and I think it would upset our guests. Can you remove it? The problem is that they all looked alike, and I have forgotten their actual positions. I only remember that there are 7 pear pies in a row, followed by 7 apple pies in a row, and finally the pomegranate pie, in clockwise order. You may taste some of them, but 3 at the most. They would be served to you, Triptie and me."

I could not think of a way, and sought help from Silenus. He picked out the pomegranate pie all right, and we avoided an awkward situation.

That afternoon, Zeus arrived at Eleusis to conduct arbitration. In the end, an agreement was hammered out. Persephone would marry Hades and rule Tartarus jointly with him for 7 months per year, one month for each pomegranate she had eaten. She could spend the other 5 months above ground. Zeus and Hades departed after Demeter had lifted her injunction.

The next morning, Demeter gave Triptie a magical bag containing an unlimited amount of barley seeds, a plough and a chariot. She sent him off to travel all over Greece, teaching people to sow barley seeds and reap the harvest. She also invited me as her special guest to visit Mount Olympus for a month.

Our parents gave their approval eagerly. With both boys gone, my mother took over temporary care of the royal cows. My father gave Silenus a one-month vacation and asked him to accompany me at least as far as the foothills of Mount Olympus.

Upon arrival at our destination, he and I checked into the Mount Olympus Inn at Demeter's expense. She said that I had to stay there for now. There were some formalities before I could be granted a day pass into the Mount Olympus Palace.

Since there was nothing to do, I asked Silenus how he picked out the pomegranate pie.

Solution

"We only know the combinations of the fruit pies and their relative positions, sir," I remarked. "How did you get started?"

"What do you think, Eubie?"

"In the absence of any explicit clues," I said, "the only thing you can do is to taste any one of the pies."

"Correct. Let us number the pies 1 to 15 in clockwise order, and call the one I tasted number 1. If it was the pomegranate pie, then I had succeeded already."

"You wouldn't be that lucky. Suppose for argument's sake that number 1 was a pear pie. Let me see. There are 7 cases. I am going to list all of them in a chart. I will use P for a pear pie, A for an apple pie and G for the pomegranate pie."

Pie	Pie Type in Case						
Number	1	2	3	4	5	6	7
1	P	P	P	P	P	P	P
2	P	P	P	P	P	P	A
3	P	P	P	P	P	A	A
4	P	P	P	P	A	A	A
5	P	P	P	A	A	A	A
6	P	P	A	A	A	A	A
7	P	A	A	A	A	A	A
8	A	A	A	A	A	A	A
9	A	A	A	A	A	A	G
10	A	A	A	A	A	G	P
11	A	A	A	A	G	P	P
12	A	A	A	G	P	P	P
13	A	A	G	P	P	P	P
14	A	G	P	P	P	P	P
15	G	P	P	P	P	P	P

"You see that I would know that number 8 was an apple pie," said Silenus. "What else do you notice?"

"The pomegranate pie was one of numbers 9 to 15."

"Which pie would you taste next?"

"I am beginning to see your idea. I would taste number 12. If it was the pomegranate pie, all would be well. Suppose it was a pear pie. Then we were in Cases 1 to 3. Now I would taste number 14. If it was the pomegranate pie, all would be well. If it was a pear pie, then number 15 was the pomegranate pie. If it was an apple pie, then number 13 was the pomegranate pie."

"Excellent! What if number 12 was an apple pie?"

"Then we were in Cases 5 to 7. Now I would taste number 10. If it was the pomegranate pie, all would be well. If it was a pear pie, then number 11 was the pomegranate pie. If it was an apple pie, then number 9 was the pomegranate pie. Of course, number 12 might have been an apple pie, or number 1 initially. However, the analysis in each case would be analogous."

Chapter 10

The Palace Admittance Questionnaire

The next morning, I was up before Silenus, and went down to the lobby of the inn.

A messenger boy of my age came up to me and asked, "Am I addressing His Highness, Prince Eubuleus of Eleusis?"

I had never been called that before. When I went out to feed the pigs, I was in my short tunic and bare feet. I was wearing a pallium and sandals now, but still did not project any regal image.

"I suppose so," I answered hesitantly.

"The goddess Demeter has requested a Mount Olympus Palace day pass for you. You have to come with me to the Mount Olympus Bureau to fill in a questionnaire."

The bureau was a short walk away. A clerk explained to me that the Palace Admittance Questionnaire consisted of 9 questions, to be answered "Yes" or "No". Each answer was either acceptable or unacceptable. To get the day pass, all my answers had to be acceptable.

He took me to a room, sat me down at a table and gave me the questionnaire. I was nervous. I began to fear that my dream holiday might be nipped in the bud. I was surprised and frustrated to find the questionnaire written in Egyptian hieroglyphs, of which I knew nothing. At the end of each question were two words separated by a slash. I could only guess that they were "Yes" and "No".

Unable to read the questions, I could not even know whether I was giving true answers or not. Since I could do nothing about it, I put the moral issue aside and hoped that I could get the day pass somehow. Since they were likely to ask whether I had some kind of disease or had committed some crimes, I decided to circle the answer "No" for every question.

After I handed in the completed questionnaire, the clerk said he had both good news and bad news. The bad news was that only 5 of my answers were acceptable. The good news was that I could come back the next day, and fill out the same questionnaire again. In fact, I could do so once a day until all my answers were acceptable.

When I got back to the inn, Silenus was up. I told him all about it.

"I don't know much about Egyptian hieroglyphs either, Eubie," he said. "Don't worry. I am pretty sure they are not trying to keep you away from the palace. Perhaps they are just buying time, so that they can investigate your background thoroughly, rather than going by what Demeter may have said."

"Since there are 2 possible answers to each of the 9 questions, sir, the number of combinations is $2^9 = 512$. It will take me almost a year and a half to try them all, but I only have one month."

"You don't have to try them all. Knowing the result of each of your attempts, you can get past the questionnaire in at most 9 more days."

I thought for a while, and came up with an idea.

"I will change exactly one of the answers, taken in order, on each of the following 8 days. My score will tell me whether changing the answer on that day is a good move or not. Moreover, since I now know what answers were acceptable for the first 8 questions, I can deduce whether my answer to the last question is acceptable or not from my initial score. Therefore, I can ace the questionnaire on day 10."

"I think you can cut the waiting period down to 8 days. While you are enjoying your visit to the palace, I will look up some old friends around this area. I will come back at the end of the month to take you back to your parents."

"Why don't you come with me to the palace, at least for a few days? Once I can ace the questionnaire, so can you."

"No, I have not been invited, and may not even be allowed to fill in a questionnaire. In any case, Apollo is in the palace. He is the God of Music, Poetry, Medicine, Archery and Unmarried Men. He chooses to appear as a young man, even though he is my grandfather. He is every bit as immature as he looks. When my father was born, Apollo took him to the palace so that the other gods and goddesses could laugh about how ugly the newborn baby was. I have sworn that I will never go there."

The memory upset Silenus so much that I left him alone for the rest of the day. I tried my best to find a strategy that would reduce my waiting period to 8 days.

Solution

I started with a simpler problem, reducing the number of questions from 9 to 5. Using my simple strategy, the waiting period would be 6 days. After much effort, I came up with a strategy which saved me 1 day.

There was no better way to start than answering "No" to all 5 questions on day 1. If my score was 0 or 5, all was well.

Suppose my score was 1 or 4. By symmetry, it might be taken to be 4. There were 5 possible cases for the acceptable answers.

Case	# 1	# 2	# 3	# 4	# 5
1	Yes	No	No	No	No
2	No	Yes	No	No	No
3	No	No	Yes	No	No
4	No	No	No	Yes	No
5	No	No	No	No	Yes

In the next 3 days, I would answer the questions as shown in the following chart. My answer to the last question was unchanged.

Day	# 1	# 2	# 3	# 4	# 5
2	Yes	Yes	No	No	No
3	No	Yes	Yes	Yes	No
4	Yes	No	Yes	No	No

My scores for these three days would be as shown in the following chart.

Day	Case 1	2	3	4	5
2	4	4	2	2	2
3	1	3	3	3	1
4	4	2	4	2	2

Since the 5 triples were distinct, I could tell which case I was in, and ace the questionnaire on day 5.

Suppose my score on the first day was 2 or 3. By symmetry again, it might be taken to be 3. This time, there were 10 possible cases.

Case	# 1	# 2	# 3	# 4	# 5
1	Yes	Yes	No	No	No
2	Yes	No	Yes	No	No
3	Yes	No	No	Yes	No
4	Yes	No	No	No	Yes
5	No	Yes	Yes	No	No
6	No	Yes	No	Yes	No
7	No	Yes	No	No	Yes
8	No	No	Yes	Yes	No
9	No	No	Yes	No	Yes
10	No	No	No	Yes	Yes

In the next 3 days, my answers were exactly the same as above. My scores would then be as shown in the following chart.

Day	Case									
	1	2	3	4	5	6	7	8	9	10
2	5	3	3	3	3	3	3	1	1	1
3	2	2	2	0	4	4	2	4	2	2
4	3	5	3	3	3	1	1	3	3	1

These 10 triples were also distinct, and the analysis was now complete.

I now used the techniques developed in the solution to the simplified problem on the original problem.

Again, I began with answering "No" to all 9 questions on day 1. On days 2 to 4, I changed only answers in questions 1 to 4 as before. On days 5 to 7, I changed only answers in questions 5 to 8 in an analogous manner.

Day	# 1	# 2	# 3	# 4	# 5	# 6	# 7	# 8	# 9
2	Yes	Yes	No	No	No	No	No	No	No
3	No	Yes	Yes	Yes	No	No	No	No	No
4	Yes	No	Yes	No	No	No	No	No	No
5	No	No	No	No	Yes	Yes	No	No	No
6	No	No	No	No	No	Yes	Yes	Yes	No
7	No	No	No	No	Yes	No	Yes	No	No

From my scores on days 2 to 4, I could determine the acceptable answers to questions 1 to 4. From my scores on days 5 to 7, I could determine the acceptable answers to questions 5 to 8. From my score on day 1, I could deduce the acceptable answer to question 9. Thus I could ace the questionnaire on day 8.

Chapter 11

The Midas Touch

I put my plan into action the next day. That did not take long, and I was left with nothing to do for the rest of the day. I had nowhere to go either. Much of the foothills of Mount Olympus were covered by the Mount Olympus Cemetery. Many famous Greek people were buried there, but I was not interested in them because they could not talk back. So I pestered Silenus for a story.

"Have you heard of the Midas touch, Eubie?" he asked.

"No, sir," I replied. "What is it?"

"Midas used to be the King of Macedonia. He planted the first rose garden in the world, and spent all his days feasting and listening to music. You know that I was with Dionysus in India. Being the inventor of wine, he held a drinking party every night. On the way home, his army passed through Macedonia. That night, I got more drunk than usual, and woke up the next morning as a prisoner of Midas. His gardeners had found me entangled in his best rose bush."

"He must have been very angry," I said.

"He was, but I placated him by telling exciting stories about India. Enchanted, he kept me with him and sent a messenger to catch up with Dionysus. Dionysus had moved on without noticing my absence. He was pleased to learn that I was safe. He agreed to release me from his service. He also agreed to grant Midas one wish."

"I guess he must have asked for the Midas touch, whatever that was."

"You are right. Everything he touched would turn into gold. At first, it was just great fun, making golden roses out of those in his garden. Then he wanted to use his new power for gain."

"With an unlimited supply of gold, he could do almost anything."

"He wanted to take over the more affluent Kingdom of Phrygia. It was a democracy. An election was to be held between the current King and a challenger. There would be three rounds. The first round was held in the lower house, which had 349 members. The second round was held in the upper house, which had 249 members. The third round was held in a joint session of both houses. To succeed, a candidate had to win the majority of votes in the first two rounds. In the third round, he had to win the majority of votes among the members of the lower house as well as among the members of the upper house."

"Why did they need a third round? Wouldn't that be a foregone conclusion?"

"Members often changed their votes."

"If Phrygia was a true democracy, the Midas touch wouldn't influence the election, would it?"

"Unfortunately, Phrygia was a corrupt democracy. All house members were open to bribery. In fact, the number of gold coins necessary to bribe a member of the lower house was fixed and known. This was also true for the upper house, though the fixed number of coins was higher than that for the lower house."

"Couldn't Midas bribe them all?"

"His power was still weak. It would have taken him too long to make that many gold coins. He spent enough to get nominated as the challenger, and sent me to Phrygia as his campaign manager. He gave me enough gold coins to bribe 175 members in the lower house and 125 members in the upper house, plus an extra 100 gold coins for any unforeseen expenses. I had suggested that, to be on the safe side, he might consider bribing an extra member in the upper house, but he did not listen to me."

"How many gold coins were needed to bribe a member in the lower house, and how many to bribe a member in the upper house?"

"Those numbers escape my mind at the moment, but I remember that I spent more gold coins bribing 175 members in the lower house than 125 members in the upper house. Had Midas followed my suggestion to bribe 126 members in the upper house, I would have spent more gold coins there instead."

"What about the third round?"

"I investigated and found that three bribed members in the lower house and one bribed member in the upper house were likely to change their votes. So I needed to use the extra gold coins to bribe their replacements. If the extra 100 gold coins were not enough, I would have to chip in what little I had. In that case, Midas had agreed to pay me twice what I spent of my own money."

"Did you make a profit out of this?"

"I am not telling you. Anyway, Midas won the election, and became the King of Phrygia. His power also grew stronger."

"So he lived happily ever after," I said, quoting one of the favorite phrases of Silenus.

"Not at all. The food he ate and the wine he drank were turning to gold in his mouth, so that he was dying of hunger and thirst. He had also turned his own daughter into a golden statue by accident."

"There is a catch to every blessing," I remarked. "What happened to him then?"

"He begged Dionysus to take away his power and restore his daughter to him. For once, Dionysus relented and brought the girl back to life. He also told Midas that he could wash off the Midas touch in the Phrygian river Pactolus. Ever since then, people went looking for gold dust on the river banks."

"Well, it is still a good ending for him."

"It was for me too. Since I helped him buy Phrygia, he did not want me around in case I revealed the truth, inadvertently or on purpose. So he released me from his service. I had grown tired of him in any case. So I came to Eleusis and was hired by your father."

"The best ending of the story was for me and Triptie. I am so glad you are our tutor."

It was lunch time. As usual, we had our meal in the dining room of the inn. While Silenus took an afternoon nap, I tried to figure out whether he had made a profit or not.

Solution

I needed to know the number of coins necessary to bribe a member of the lower house, and the number of coins necessary to bribe a member of the upper house. Wanting to be discrete even in this hypothetical rerun, I put the former amount inside a bag and the latter amount inside a box. All bags contained the same number of coins, as did all boxes. I knew that the number of coins inside a bag was less than the number inside a box.

From the conditions given by Silenus, there were more coins inside 175 bags than inside 125 boxes. Canceling 25, this meant that there were more coins inside 7 bags than inside 5 boxes. Also, there were more coins inside 126 boxes than inside 175 bags. Canceling 7, this meant that there were more coins inside 18 boxes than inside 25 bags. I was certain that these cancellations were important steps.

However, I was unable to make any progress beyond that. I had two inequalities. I tried to combine them in a way which would not take me back to where I started. Multiplying by 18, I knew that there were more coins inside 126 bags than inside 90 boxes. Multiplying by 5, I knew that there were more coins inside 90 boxes than inside 125 bags. All I could deduce was that there were more coins inside 126 bags than inside 125 bags!

Then Silenus rejoined me and made a crucial suggestion. I had to use the fact that each bag and each box contained a *whole* number of coins.

The number of coins inside 7 bags was at least the number of coins inside 5 boxes plus 1. Hence the number of coins inside 175 bags was at least the number of coins inside 125 boxes plus 25.

Similarly, the number of coins inside 126 boxes is at least the number of coins inside 175 bags plus 7. In other words, the number of coins inside 175 bags was at most the number of coins inside 126 boxes minus 25.

Combining these two inequalities, the number of coins inside 126 boxes minus 25 is at least the number of coins inside 125 boxes plus 7. Hence the number of coins inside a box was at least $25 + 7 = 32$.

The number of coins inside 18 boxes was at least the number of coins inside 25 bags plus 1. Hence the number of coins inside 90 boxes was at least the number of coins inside 125 bags plus 5.

Similarly, the number coins inside 126 bags was at least the number of coins inside 90 boxes plus 18. In other words, the number of coins inside 90 boxes was at most the number of coins inside 126 bags minus 18.

Combining these two inequalities, the number of coins inside 126 bags minus 18 is at least the number of coins inside 125 bags plus 5. Hence the number of coins inside a bag was at least $18 + 5 = 23$.

To bribe one member of the upper house and three members of the lower house would take at least $32 + 3 \times 23 = 101$ coins. Hence Silenus had to chip in at least one, and consequently made a profit from his loan.

Chapter 12

The Collapsing Beams

At lunch the next day, Silenus and I met a young couple.

"I am Bias of Pylus," the young man introduced himself and his companion, "and this is my cousin Pero. We are getting married in this sacred place."

"Congratulations," said Silenus. "I wish you marital bliss."

"We owe our happiness to Melampus, the twin brother of Bias. My father would not let Bias marry me unless Bias could give him a herd of cows that belonged to Phylacus, a powerful lord. Phylacus refused to sell at any price, and Bias nearly died of disappointment."

"Melampus is the best of brothers," said Bias. "Once he prevented our servant from killing a brood of little snakes. In gratitude, they licked his ears while he was asleep. When he woke up, he found that he had gained the ability to understand the language of birds and insects. One day, he heard two cranes talking about me and Pero. The birds said that if I stole the cows, I would be killed by Phylacus. If anyone else did, he would just go to prison for a year."

"So Melampus stole the cows," said Pero, "got caught by Phylacus, and was put into a special prison."

"What was so special about it?" I asked.

Bias explained, "It was an 8×8 underground prison. My brother was lowered into one of the 64 rooms from above. A slab of stone, supported by a wooden beam, covered up the opening. Each room had a white flag or a black flag on its slab, and rooms separated by a wall had flags of different colors. The room at the southwest corner had a black flag."

"Ten nights before the end of his year as a prisoner," continued Pero, "Melampus heard two wood-worms talking in the beam above his head. One remarked that the head wood-worm had decided that some of them would eat at the wood really hard throughout the night, so that the beams of all the rooms with white flags would collapse at dawn the next day."

"Was he in a room with a white flag?" Silenus asked.

"He could not see the flag. He just hollered and demanded to be taken away from the prison," said Bias. "Phylacus thought that it was a trick. He said he would only allow Melampus to trade rooms with a slave-woman imprisoned in a room separated from his by a wall."

"That wouldn't do Melampus any good, would it?" I asked.

"No," said Pero. "Not knowing the color of the flag of his room, he might be trading into his own death."

"He prayed to the goddess Hera, who sent a vulture to deliver a message," said Bias. "He was to draw two diagrams of the prison. On each, he would mark off a connected part of it, consisting of one or more rooms. Hera would then tell him whether his room was inside or outside the marked part."

"Very interesting," remarked Silenus. "He must then be able to deduce the color of the flag of his room.

"Melampus thought hard and fast," Bias continued. "He came up with the diagrams. He then deduced that he was in a room with a white flag, and accepted the trade. At dawn, the beams of all the rooms with white flags collapsed. In the room where my brother was before the trade, the slabs fell down and killed the slave-woman."

"Phylacus was astonished," said Pero. "He believed that Melampus was a prophet, apologized to him and gave him the cows. He gave them to Bias, and Bias gave them to my father. My father gave me to Bias, and all ended happily."

"How did Melampus figure out that he was in a room with a white flag?" Silenus asked me.

"Let me start with a 2 × 2 prison. I call a room with a white flag a white room, and represent it with a blank square. I call a room with a black flag a black room, and represent it with a shaded square. Aha! There is a trivial solution, as shown in the diagram below."

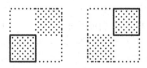

"I see," said Silenus, "but this idea isn't useful at all for larger prisons, is it?"

Solution

While Silenus was taking his afternoon nap, I continued to work out how Melampus discovered that he was in a white room. I looked for a different approach, and came up with the diagrams below.

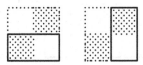

If the one response was "inside" and the other "outside", Melampus was in a black room. If the two responses were the same, he was in a white room.

I now moved onto a 4 × 4 prison. I could get 5 white rooms joined corner to corner without surrounding any black room. They would be inside the marked part in each diagram. The remaining 3 white rooms would be outside both marked parts. We now joined these 5 white rooms into a connected part in two ways, using two disjoint sets of black rooms, as shown in the diagram below.

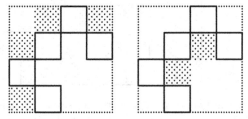

The actual diagrams were shown below. They had been expanded so that each included one of two black rooms near the southeast corner. If the two responses were the same, then Melampus was in a white room. If the responses were different, Melampus was in a black room.

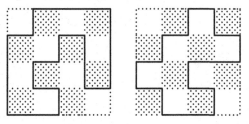

I then made this construction work for an 8 × 8 prison.

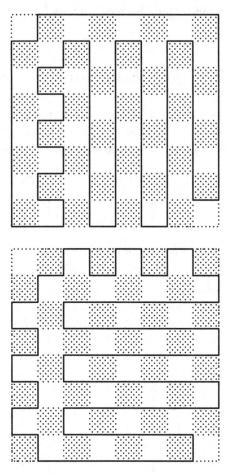

As before, if the two responses were the same, then Melampus was in a white room. If the responses were different, Melampus was in a black room.

Chapter 13

The Seating Plan

The wedding banquet was the night before. There was wine, and once again Silenus was drunk. He slept in this morning. When I came back from the Mount Olympus Bureau, I decided to stay with him to make sure that he did not get into mischief.

The following morning, I filled in the Palace Admittance Questionnaire for the fifth time. I lucked out, and aced it.

At lunch time, a huge man with lots of body hair came into the dining room. He was wearing a lion skin. He spotted us and came over.

"Silenus! You old goat!" he exclaimed. "I heard that you had perished in India."

"The reports of my death are exaggerated, my old friend," Silenus said as he stood up and embraced the stranger. "What have you been up to?"

"Actually, I died recently, poisoned by my former wife, unintentionally. She committed suicide. My father raised me from the dead and made me immortal. He also gave me a new wife and a new job. Do you remember my half-sister Hebe?"

"Yes, the Goddess of Youth. So you are married to her. Congratulations," Silenus said for the second time in two days, "I wish you marital bliss. Isn't Hera her mother? How are you getting along with your new mother-in-law?"

"I try to stay out of her way. I am not the only one with mother-in-law issues."

"What is the job?"

"I am now the official porter of the Mount Olympus Palace."

"Isn't that rather demeaning for you?"

"Not at all. I had had enough adventures in my former life. I want to enjoy some peace and quiet with Hebe."

I had been ignored so far. By then, I had deduced the identity of the stranger. The lion skin was a major clue. I decided to cut in.

"Sir, are you the mighty hero Heracles?"

"Gods are forever," he said, "but heroes are only until the next ones come along. Around here, I am just the *hairy porter*. Who may you be?"

I was too overawed for more words.

"I am Eubie's tutor," Silenus came to my rescue. "Your aunt Demeter had invited him to visit the palace."

"So you are the little boy I am looking for. My aunt has sent me to fetch you, now that you have got your day pass. It is good for three days. It will then be upgraded to a full pass, which allows you to stay overnight in the palace."

I followed Heracles up Mount Olympus. The palace sat well above the usual level of clouds at the top. It was an enormous structure with strong walls constructed by the cyclopes. The first building inside was the gatehouse. A woman was waiting in front.

"You must be Eubie," she said with a warm hug. "Welcome to our humble home. I am Heracles's wife Hebe. Aunt Demeter has told me so much about you. She is with cousin Persephone. You know about her ordeal in Tartarus. It is quite upsetting. They will see you later. Meanwhile, Heracles has to do some odd chores. Why don't you tag along so that he can show you around?"

"Sure, and I can make myself useful in some small way."

Heracles took me up to the rampart above the gate, a great vantage point.

"We are at the south end of the palace, looking towards the famous Greek cities of Athens, Delphi, Thebes, Sparta, Corinth, Argos and Mycenae. Immediately behind the gatehouse is the grandest building inside the palace, the Council Chamber, where the Olympians meet to discuss mortal affairs. Let us go there because I have to tidy it up."

The chamber had two thrones at the far end, facing the door, an enormous one with seven steps leading up to it, and a smaller one with three steps leading up to it.

"I guess these are the thrones of King Zeus and Queen Hera."

"Correct," replied Heracles. "Do you see the ten thrones along the two sides of the hall?"

"Yes, I do."

"They are for my uncle Poseidon, my aunt Demeter and my half-brothers and half-sisters."

"Who are they?"

"My father has three offspring with Hera. Hebe is the youngest. She and my aunt Hestia used to have council seats, but not any more. The other two are boys, Hephaestus and Ares. Like me, the rest are all born of my father's extra-marital affairs, the girls Athene, Aphrodite and Artemis and the boys Apollo, Hermes and Dionysus."

"The place seems rather tidy already."

"The trouble is that my father, who is always the first to arrive, is sometimes absent-minded. On those occasions, he invariably takes Hera's throne, which is the next one in the clockwise direction from his own. One at a time, the others arrive, and occupy their correct thrones unless already occupied. In such a case, the first vacant throne in the clockwise direction will be occupied instead. Today, not all of them are in the right thrones. They all leave bits of their personal belongings behind, and it is my job to sort them out and restore them to the correct thrones of the owners."

"If I am to help, I need to know the seating plan."

"It is on the inside of the door."

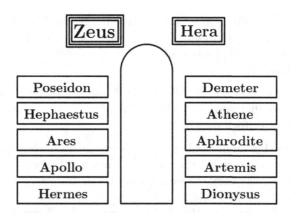

We got everything sorted out in the end. Heracles had some heavier chores to do, and sent me back to the gatehouse. Hebe was out.

While waiting, my mind wandered back to the chamber. Different incorrect seating arrangements resulted from different orders in which the others arrived after Zeus had taken Hera's throne, though different orders might lead to the same arrangement. I wondered how many possible scenarios there could be.

Solution

Since 12 gods and goddesses were too many to handle all at once, I started with the simplest scenario when the council consisted only of Zeus and Hera. When Zeus took the wrong throne, Hera had to do likewise. Hence there was only 1 scenario.

I added Demeter to the council. If she arrived last, she would get Zeus's throne since her own throne would have been occupied by Hera. Otherwise, she would get her own throne, followed by Hera taking Zeus's throne. Hence there were 2 scenarios.

I then added Poseidon to the council. I drew up the following chart, which showed that the number of scenarios had grown to 4.

Order of Arrival			Throne Occupancy		
First	Second	Third	Demeter's	Poseidon's	Zeus's
Hera	Demeter	Poseidon	Hera	Demeter	Poseidon
Demeter	Hera	Poseidon	Demeter	Hera	Poseidon
Hera	Poseidon	Demeter	Hera	Poseidon	Demeter
Poseidon	Hera	Demeter	Hera	Poseidon	Demeter
Demeter	Poseidon	Hera	Demeter	Poseidon	Hera
Poseidon	Demeter	Hera	Demeter	Poseidon	Hera

There were 6 different orders of arrival. In the 2 cases when Poseidon arrived last, he had to occupy Zeus's throne. In the other 4 cases, there were at least 2 unoccupied thrones, including his and Zeus's, so that he would get his own throne.

The answers being 1, 2 and 4 so far, it was reasonable to expect that the addition of another god or goddess would double the answer. To test this, I added another council member.

As before, Poseidon would get Zeus's throne if he were the last one to arrive, and his own throne otherwise. In the second case, if we removed him and his throne, the answer would be 4. In the first case, if we removed him and his throne while moving whoever was on his throne over to Zeus's, the answer would again be 4. Hence there were $4 + 4 = 8$ scenarios as expected.

This argument might be repeated by adding the other members 1 at a time. Since there were 11 other gods and goddesses apart from Zeus, the number of scenarios would be $2^{11-1} = 1024$.

Chapter 14

The Danaïds

I had dozed off when Hebe came back to the gatehouse.

She woke me and said, "I have just been over to help Aunt Demeter prepare dinner. She has invited you and me. Heracles will join us after work. It is almost time now."

The private quarters of Zeus and of Hera were directly behind the Council Chamber. Behind them were the private quarters of the other ten council members. When we got to Demeter's, Heracles was already there.

"Eubie," said Persephone, "I don't think I had a chance before to thank you for letting my mother know what happened to me. I want you to know that I am deeply grateful to you."

"Think nothing of it, Your Majesty," I said humbly.

"Please don't call me Your Majesty, or I will call you Your Highness."

"But I can't call you *Phonie*," I said.

Demeter laughed and said, "We call her Persie. Let's be seated at the table. Hebe will be serving dinner."

Dinner began with wild mushroom soup. The main course was ambrosia, Food of the Gods. It was an uncooked mixture of honey, water, fruit, olive oil, cheese and barley. For drinks, we had nectar made from fermented honey.

"I have been living it up since coming back from Tartarus, Eubie," Persephone said. "It is really boring down there. Fortunately, I have a good friend. Hypermnestra is the youngest daughter of King Danaus of Argos."

"If she is in Tartarus, Persie," I said, "wouldn't she be a bad person?"

"It is a common belief that only bad people go to Tartarus when they die, but this is not correct. All dead people go there. It is only then that their past lives would be judged as good or bad. Hypermnestra happens to be very nice. She even tried to help her sisters escape."

"From Tartarus?" exclaimed Hebe. "Tell us more about it."

"Danaus had fifty daughters. His brother Aegyptus had fifty sons, and forced him to marry his daughters to their cousins. Danaus told the Danaïds to stab their husbands' hearts on the common wedding night. Only Hypermnestra disobeyed since she had fallen in love with her husband. Her sisters were not judged too harshly because they had only obeyed their father's orders."

"Did they succeed in escaping?" I asked.

"Unfortunately, no," answered Persephone.

"Why not?" asked Hebe.

"Hypermnestra found a boat for them to cross the River Styx, which separated Tartarus from the rest of the world. It would require two of them to row and was large enough to carry an additional passenger. However, her sisters were very snobbish. They ranked themselves by age from 1 to 49. Two Danaïds whose ranks differed by more than 1 refused to be in the boat at the same time."

"So they were faced with a mission impossible," Demeter remarked, "because at most two of them could be in the boat at a time, and the same two who crossed over had to bring the boat back themselves."

"Hypermnestra pleaded with her sisters to no avail," said Persephone. "Then she offered to escape with them, provided that she was given the rank 1 as well. This her sisters also refused. So she gave up."

It was getting late. Heracles took me back to the Mount Olympus Inn. I told Silenus all about my exciting experience that day.

"About this attempted escape, Eubie," asked Silenus, "do you think it was possible had the sisters agreed to Hypermnestra's condition?"

"I don't know off-hand, but she sounds like someone who knows what she was doing. I would place a small bet that she had a plan."

"It would not work because the attempt would be discovered long before its completion. Can you determine the minimum number of crossings this must take?"

Solution

I was tired, and went to sleep. In the middle of the night, I woke up as I had been working subconsciously on the problem. I first considered the special case where only the sisters ranked 1 and 2 tried to escape with Hypermnestra, also ranked 1. Then they could escape together in 1 crossing.

Now I added the sister ranked 3. The escape could be done in 5 crossings, as shown in the following chart.

Crossing Number	Ranks of Sisters		
	on Near Bank	in Boat	on Far Bank
1	3	1,1,2	1,1,2
2	1,2,3	1,2	1
3	1	2,3	1,2,3
4	1,1,2	1,2	3
5		1,1,2	1,1,2,3

The odd-numbered crossings must be from the near bank and the even-numbered ones from the far bank. Since the first and the last crossings are both from the near bank, the number of crossings must be odd.

Next I added the sister ranked 4. The escape could be done in 9 crossings, as shown in the following chart below.

Crossing Number	Ranks of Sisters		
	on Near Bank	in Boat	on Far Bank
1	3,4	1,1,2	1,1,2
2	1,2,3,4	1,2	1
3	1,4	2,3	1,2,3
4	1,1,2,4	1,2	3
5	4	1,1,2	1,1,2,3
6	2,3,4	2,3	1,1
7	2	3,4	1,1,3,4
8	1,1,2	1,1	3,4
9		1,1,2	1,1,2,3,4

The first 4 crossings were essentially the same as before, with the sister ranked 4 observing the proceeding on the near bank. The next 4 crossings were used to get her over to the far bank in exchange for the sisters ranked 1, 1 and 2. These three could escape in crossing number 9.

It would appear that each additional sister would required 4 more crossings. If so, the escape could be done in $9 + 4(49 - 4) = 189$ crossings. I added the sister ranked 5 to see if my idea was correct.

| Crossing | Ranks of Sisters | | |
Number	on Near Bank	in Boat	on Far Bank
1	3,4,5	1,1,2	1,1,2
2	1,2,3,4,5	1,2	1
3	1,4,5	2,3	1,2,3
4	1,1,2,4,5	1,2	3
5	4,5	1,1,2	1,1,2,3
6	2,3,4,5	2,3	1,1
7	2,3	4,5	1,1,4,5
8	1,1,2,3	1,1	4,5
9	3	1,1,2	1,1,2,4,5
10	1,2,3	1,2	1,4,5
11	1	2,3	1,2,3,4,5
12	1,1,2	1,2	3,4,5
13		1,1,2	1,1,2,3,4,5

This time, the first 8 crossings were essentially the same as before, with the sister ranked 5 observing the proceeding on the near bank. The last 5 crossings were essentially the same as the case before that, with the sisters ranked 4 and 5 observing the proceedings on the far bank.

The Danaïds could have escaped, but could it be done with less than 189 crossings? There were 49 Danaïds plus Hypermnestra. They had to cross over with 3 sisters 48 times, with a net gain of 1 sister each time except that in the final crossing, the net gain would be 3. They might need as few as $2 \times 47 + 1 = 95$ crossings.

Obviously, it was impossible for three sisters to be in the boat in every crossing from the near bank, because it must be done by the sisters ranked 1, 1 and 2, and they must all come back. Since at most two of them could come back right away, it would take 2 more crossings to bring all of them back. So there could be 3 sisters in the boat in every other crossing from the near bank. This meant that there must be at least 3 other crossings between 2 crossings from the near bank with 3 sisters. So the minimum number of crossings required was indeed $4 \times 47 + 1 = 189$.

Chapter 15

The Quarrelsome Residents

I slept in the next morning until Silenus woke me up.

"Heracles is waiting to take you up Mount Olympus, you lazy bones."

I scrambled out of bed, put on my pallium and sandals, and ran down to the lobby. I apologized to Heracles. Seeing that I was still tired, he picked me up, carried me on his shoulders and took me up Mount Olympus.

"Today we are at the north end of the palace, Eubie," said Heracles. "Here, you will find the kitchen, the banquet hall, the armory, the workshops and the servants' quarters. Then there are sheds for chariots, stables for horses, kennels for hounds, and a sort of zoo where the Olympians kept their sacred animals. Behind all these is the room of the three Fates, named Clotho, Lachesis and Atropos. They are the oldest goddesses in existence, too old for anybody to remember where they came from. For each mortal, the Fates spin a linen thread to measure the length of the person's life, and snip it off when the time comes."

Separating the north end from the south end of the palace was a row of houses, stretching as far as the eye could see in both directions, way up into the sky.

"What is that, sir?" I asked.

"This is the Creamy Way, where the minor gods and goddesses live. There are far too many of them. Hebe moved here after she resigned from the council, until she married me. My aunt Hestia is still here. Others are Dionysus's mother Semele, a mortal woman whom my father turned into a goddess; Poseidon's wife Amphitrite; and the Nine Muses who sing in the banquet hall. A particularly interesting one is Eros, the God of Love. He is the son of Aphrodite, the Goddess of Love and Beauty. The wicked little boy enjoys shooting arrows at people to make them fall ridiculously in love. He is moving today, and I think you may enjoy meeting him and giving him a hand."

"Is he being promoted to the council or kicked out of the palace?"

"Neither," said Heracles. "The residents on the Creamy Way are the most quarrelsome people you can imagine. That is why they all live alone. If two of them happen to live in adjacent houses, they want to move away from each other to the houses on the opposite sides the next day. Sometimes, this puts two of them in the same house, and they will move off in opposite directions the day after. Of course, I have to help with their moves, and this has been going on for days and days. Will the madness ever end?"

"If they come to the ends of the row," I asked, "where would they move?"

"Unfortunately, the row is long enough so that you may consider it without ends."

Eros looked to be about my age, and very playful. We got along very well. When we had finished moving his stuff, I returned to the gatehouse for lunch. Hebe had baked some barley cakes. While we were eating, she told me all about her brothers as well as her half-brothers and half-sisters.

"Hephaestus, the God of Goldsmiths, Jewelers, Blacksmiths, Masons and Carpenters, is a skillful workman and a gentleman. At birth, he was very weak. It displeased father so much that he threw the baby over the palace walls. In his fall, Hephaestus broke a leg so badly that he was crippled for life."

"How could King Zeus do that to his own son?" I exclaimed.

"Father dislikes Hephaestus but loves Ares, the God of War. Tall, handsome, boastful and cruel, Ares is every bit as stupid as father. Try to stay out of this bully's way. He loves fighting for its own sake."

"My tutor Silenus the satyr has told me that Apollo is his grandfather."

"Apollo is just a silly brat, always in trouble with father. He had rebelled once or twice and got well punished each time. Father hopes that he will eventually learn to behave more sensibly. His twin sister Artemis, the Goddess of Hunting and Unmarried Women, actually taught him medicine and archery."

"She must be a most accomplished goddess."

"The most accomplished one has to be Athene, the Goddess of Wisdom. She is a bit of a mystery. Nobody knows who her mother is. She taught Hephaestus how to handle tools, and knew more than anyone else about pottery, weaving and all useful arts. She is also a battle-goddess, yet never goes to war unless forced, and when she fights, she always wins."

"What is Aphrodite like, the mother of my new friend Eros?"

"She is a wild girl. To keep her out of mischief, father married her to Hephaestus. Hephaestus was ecstatic, and gave her a magic belt as a wedding present. Whenever she wears it, she can make whoever she wants love her madly. She has an ongoing affair with Ares openly, and she has many other admirers. It is believed that neither Hephaestus nor Ares is the father of Eros."

"There are two more gods, Hermes and Dionysus."

"Hermes is the God of Merchants, Bankers, Thieves, Fortune-tellers and Heralds. He is father's messenger, and you heard last night that he leads the souls of dead people to Tartarus. Actually, he is very nice and very clever. He has invented the letters of the alphabet with some help from the Fates, as well as arithmetic, astronomy, musical scales, weights and measures, the arts of boxing and gymnastics."

"Wow!"

"I am deeply impressed too. That was why I gave up my council seat to him. Your tutor Silenus was present at his birth, but I will let the old satyr tell you that story himself. However, he does not know everything. Here is an excerpt from my diary. You can read it after you have talked to Silenus."

"Silenus also had some connections with Dionysus."

"Dionysus, unlike Hermes, is rather nasty. Nobody likes him. However, he is father's favorite, just because he invented wine. Father took the council seat from Aunt Hestia and gave it to him."

"Isn't Heracles coming for lunch?" I asked as we had just finished eating.

"Dear me," exclaimed Hebe. "I can't stop myself when I start talking. I have completely forgotten about my husband. Please take some barley cakes to him."

The other people moving that day were all adults. Heracles told me to visit the zoo. I saw a bear, a lion, a cow, a wild boar, a wild cat, a peacock, a crane, an eagle, an owl, a tortoise and a tank full of fish. There were also tigers, stags, white bulls, mice, swans, herons and snakes.

I went back to the Mount Olympus Inn early because it would be my last night with Silenus until a month later. I was convinced that the new labor of Heracles, about moving the quarrelsome residents, had to come to an end. Silenus helped me find a proof.

Solution

"While the number of houses may be regarded as infinite, Eubie," he said, "the number of residents is finite, albeit very large. Let us number the houses $\ldots, -2, -1, 0, 1, 2, \ldots$ and list the tenants according to their current addresses from the houses with lower numbers on the left to the houses with higher numbers to the right. This order is unchanged after every move. The leftmost tenant never moves to the right while the rightmost tenant never moves to the left."

"I notice another thing which is unchanged after every move, sir," I said, "the sum of the numbers of all occupied houses. The loss of 1 by the resident moving to the left and the gain of 1 by the resident moving to the right cancel each other."

"Amidst the continual changes, this fixed value can serve as a reference framework. However, it may be better to find another value, which changes in a known pattern."

"I have an idea. Let us replace the number of an occupied house by the square of that number. Call the sum of these squares the magic number. Suppose two residents in the same house move. Now the difference between consecutive squares is always an odd number, and this difference always increases by 2. It follows that the magic number increases by the difference of two odd numbers in a row, namely 2."

I worked out an example. If both residents started in number 7, the old magic number would be $7^2 + 7^2 = 98$ while the new magic number would be $6^2 + 8^2 = 100$. The difference was indeed 2.

"What happens if they live instead in adjacent houses?"

I could not see the pattern. I worked out a couple of examples aloud.

"Suppose they start in numbers 3 and 4. Then the old magic number is $3^2 + 4^2 = 25$ and the new magic number is $2^2 + 5^2 = 29$. Suppose they start in numbers 7 and 8. Then the old magic number is $7^2 + 8^2 = 113$ and the new magic number is $6^2 + 9^2 = 117$. The difference is 4 in both cases."

"It will always be 4. Three differences of squares are involved. In your first example, they are $3^2 - 2^2 = 5$, $4^2 - 3^2 = 7$ and $5^2 - 4^2 = 9$. In your second example, they are $7^2 - 6^2 = 13$, $8^2 - 7^2 = 15$ and $9^2 - 8^2 = 17$. Now 5, 7 and 9 are three odd numbers in a row, as are 13, 15 and 17. The difference between the first and the last is always 4."

"So the magic number always increases, either by 2 or by 4. If the moving never ceases, then the magic number will increase without limit. This can only happen if the leftmost resident keeps moving to the left while the rightmost resident keeps moving to the right. This is clearly impossible."

"Indeed, and I can give you a formal argument if you wish."

"No, thank you. I am probably not ready for that in any case. I am convinced that Heracles's ordeal must come to an end some day if there are no new residents."

Chapter 16

The Bears in the Caves

During dinner, I asked Silenus about the birth of Hermes.

"Once upon a time," began Silenus in his usual way, "when I was still young, I met Apollo for the first time. I did not know then that he was my grandfather, especially since he did not look any older than I. He had just lost a fine herd of white cows in Arcadia, and offered a reward for the discovery of the thief. I was living in the area, and went with eleven friends, all young satyrs, to join the search. We found some strange tracks, and deduced that they came from cows wearing shoes made of bark and plaited grass."

"A very clever thief," I said.

"We followed the tracks to a mountain. The sun was setting, and the air was cold. So we decided to spend the night in caves. Each of us wanted his own cave. We found 49 of them. Then one of us who had been to that place remembered that there were 7 bears who lived there, but he could not remember which caves they were in. I came up with a scheme. The first 48 caves were partitioned equally among the twelve of us. Each friend would try the caves on his list one at a time, and move into the first one not occupied by a bear. If there were bears in every cave on his list, he would take the last cave, which was unassigned."

"What happened if each of two friends found bears in every cave on his list?"

"This could not happen, as there were only 7 bears. The next morning, one of the caves caved in. Fortunately, none of us was inside, nor was any bear. However, had that happened the day before, I would not have been able to come up with a working scheme, for the simple reason that none could exist."

"Why?"

"I will leave that to you. We soon came across a young woman filling a pitcher at a stream. We secretly followed her up the mountain, and saw her disappear into a cave above our row of forty-nine caves. We heard wonderful music coming from inside."

"Was she some goddess?"

"It turned out that she was a nurse. With her was a woman who was sleeping. The musician was a baby boy, playing on a musical instrument made by stringing cow-guts tight across the hollow shell of a tortoise. I noticed a newly flayed white cowhide hung up to dry, and accused the boy of stealing Apollo's white cows."

"I assume that the boy must have been Hermes."

"Yes, he was born only the day before, but being a god, he grew within a few minutes to the size of a four-year-old. The nurse accused me of laying false charges on an innocent child, and told me to keep my voice down lest I should wake the mother."

"How was the matter resolved?"

"At that moment Apollo flew down, went straight into the cave, caught hold of Hermes, who was pretending to be asleep, and carried him to Mount Olympus."

"What happened then?"

"I didn't know. A council meeting was probably called. Hebe still had a seat then, and could tell you all about it."

Then I remembered the excerpt from her diary, and read it together with Silenus.

This morning, Apollo brought a little boy before the council and accused him of theft.

Father frowned and asked, "Who are you, little boy?"

"Your son Hermes, father," Hermes answered. "I was born yesterday."

"Then you certainly must be innocent of this crime."

"You know best, father."

*Apollo screamed, "Don't give **me** that. I wasn't born yesterday. He stole **my** cows!"*

Hermes little eyes rolled around for a while before he finally said, "I was too young to know right from wrong yesterday. Today I do, and I beg your pardon. You may have the rest of those cows, if they are yours. I killed only one to get the cow-guts."

In the afternoon, I went with them to the cave, where Hermes took the tortoise-shell lyre from under the blankets of his cradle and played so beautifully on it that Apollo exclaimed, "Hand over that instrument. I am the God of Music!"

"Very well, if I may keep your cows," said Hermes.

They shook hands on the bargain, the first ever made. Then we came back to tell father that the affair had been settled.

We were spending our last night at the Mount Olympus Inn. We would be checking out in the morning. So we started packing. I had a lot less with me, and finished much earlier than Silenus. So I sat down and tried to understand why he said that with only 48 caves, no working scheme for allotting caves could exist.

Solution

For argument's sake, I assumed that Silenus was mistaken, and there was a working scheme. In any such scheme, each friend would receive a list of numbers of caves to be tried. Since there were 7 bears and no unassigned cave, each list must be of length 8. Of course, some numbers would appear in several lists, and two friends might be directed to the same cave. They would not notice each other in the dark, but would kick up a big fuzz in the morning. The scheme would be considered not working.

It would be best to limit the overlap between lists as much as possible. With $8 \times 12 = 96$ entries, each number should appear 2 times, even though this was not required. I draw up an example based on this assumption, combining the 12 lists into a 8 × 12 table.

1	12	2	3	4	5	6	37	8	9	47	11
2	8	12	13	14	15	16	17	18	19	20	21
3	9	16	22	29	23	24	25	46	27	28	29
4	11	17	23	28	30	31	32	33	34	35	36
5	1	18	24	22	33	37	7	38	39	40	48
6	13	19	25	30	34	38	41	42	43	44	45
7	14	20	26	31	35	39	42	44	45	10	47
10	15	21	27	32	36	40	43	26	46	48	41

I now divided the argument into two cases.

Case 1. The 48 numbers in the top four rows were not distinct.

In other words, there were two copies of the same number, necessarily in different columns. In the example, the number 16 appeared as the third entry in column 3 and as the second entry in column 7. If caves 2, 6 and 12 were occupied by bears while cave 16 was not, both the third friend and the seventh friend would move into cave number 16. The lowest placing of the two identical numbers would be if both were on the fourth row. Then each had three numbers above, and there were enough bears to occupy those caves.

Case 2. The 48 numbers in the top four rows were distinct.

Any number on the fifth rows had to appear again in the first four rows, in different columns. One copy had 4 numbers above it, while the other had at most 3 numbers above it. Again, there were enough bears to occupy those caves.

When Silenus had finished his packing, I presented my proof to him. He was delighted at the progress I had been making in problem solving.

"Eubie, consider this hypothetical scenario. Suppose a cave caved in the day before, killing a bear inside. Could there have been a working scheme?"

"Let me see, sir," I said. "I think so. I would partition the 48 caves equally among the 12 friends, with no unassigned cave this time."

"What would have happened if one of us had found bears in every cave on his list?"

"Well, there were only 6 bears now. If he had found 4 of them, he could try the last caves on the lists of the others. None of these friends would be there. Since there were at most 2 bears among these caves, he would find one without a bear no later than on his third try."

Chapter 17

The Phrygian Music Festival

Silenus left early the next morning because he had an ambitious travel plan for the day. I took my satchel down to the lobby and waited for Heracles. He did not show up at the usual time. I was wondering whether I should go up the mountain by myself when Hebe came in.

"Eubie," she said, "Heracles will be coming down soon. He is escorting the nine Muses here for a funeral."

"Who died?" I asked.

"Orpheus of Thrace, son of Calliope, the Muse of Eloquence and Epic Poetry."

"Who are the other muses?" I asked. I remembered Heracles telling me that they sang in the banquet hall in the palace.

"Erato is the Muse of Romance and Romantic Poetry, Polyhymnia is the Muse of Harmony and Lyric Poetry, Terpsichore is the Muse of Dance and Light Poetry, Melpomene is the Muse of Tragic Theater. Thalia is the Muse of Comic Theater, Clio is the Muse of History, Euterpe is the Muse of Music and Urania is the Muse of Astronomy."

The party arrived in due course. Along with the nine Muses, Athene, Apollo and Hermes also came.

In the eulogy, Calliope said, "Besides being a poet, my son played the lyre very well. His greatest achievement was winning the competition in the Phrygian Music Festival a year ago. His deepest regret was losing his beautiful wife Eurydice. She died, bitten by a poisonous snake. My son boldly went down to Tartarus, playing his lyre. He got past Charon and Cerberus and charmed Hades into freeing Eurydice, on the condition he must walk in front of his wife and not look behind him until they were above ground. At the last minute, he feared that Hades might be tricking him. He forgot the condition and ...". Calliope broke down and could not finish the sentence.

No details about Orpheus's death were given. In the reception which followed, Heracles introduced me to Athene, Apollo and Hermes.

"All three of us were at the competition in Phrygia. King Midas was the grand patron of the event, and judged the competition along with his benefactor Dionysus. As the God of Music," said Apollo, "I was the first one invited."

"I had just invented the double flute," said Athene.

"I had made a new tortoise-shell lyre," said Hermes.

"Were there any other competitors?" Hebe asked.

"Apollo's son Pan was there with his pan pipes," said Athene.

"Pan's son Silenus was in the service of King Midas at the time," said Hermes. "Like most satyrs, he played the drum. Marsyas, a Phrygian shepherd who played the flute, also wanted to enter the competition."

"We couldn't let in any riff-raff," said Apollo. "I convinced Midas that with the festival lasting only 4 days, the number of competitors must be restricted to 6."

Hermes said, "The rule of the competition stipulated that on each day, some of us would perform while the others joined the audience, so that everyone had a chance to listen to the music performed by every other competitor."

We were sitting at a quiet corner. Athene looked around carefully, lowered her voice and said to me, "Eubie, what you are about to hear must be kept in the strictest confidence."

"I will not say anything to anyone else, I promise."

"Well," Athene continued, "You all know that Dionysus is unpopular among us. He is unpopular among the mortals too. Orpheus for one did not worship him, considering his drunken behavior as setting a very bad example."

"But the Grand Prize was awarded to Orpheus!" exclaimed Apollo.

"Dionysus thought he could win Orpheus over, but he didn't," said Hermes. "You may remember that Orpheus refused to accept the award from him but from Midas. Shortly after the festival, I found out that Dionysus had murdered Orpheus. He even mutilated the corpse by cutting the body into small pieces, and throwing the head into the River Pactolus. The head only washed up on the island of Lemnos recently. That is why the burial was delayed till now."

"My goodness, and Calliope suspected me of doing away with Orpheus because she thought I would be angry about losing," complained Apollo.

"You were," said Athene, "and you do have a mean streak, although you are tamer than Dionysus."

"How could he get away with this?" asked Apollo.

"He is currently very much father's favorite son," said Athene. "You, on the other hand, are quite the opposite."

After the funeral, I went up Mount Olympus with the rest.

Hebe said to me on the way, "We have talked it over, Eubie. It is more convenient for you to stay with us at the gatehouse than with two women in Aunt Demeter's place."

"This is my first choice too," I said, thrilled. "I want to be close to Heracles, and to you too."

Having been away the whole morning, Heracles had a lot of tasks on his hands. I told Hebe something of my life so far. She was particularly interested in all the puzzles Silenus had given me.

She then asked, "Was Apollo right when he claimed that a 4-day festival could accommodate only 6 competitors under the given rules?"

Solution

"Let us suppose that the festival lasts 3 days and there are 6 competitors," I said. "Then at least 2 competitors must perform on some day, say Athene and Hermes on the first day. Then they must perform for each other, say Athene for Hermes on the second day and Hermes for Athene on the third day."

"Since Hermes is in the audience only on the second day, while Athene is in the audience only on the third day," said Hebe, picking up the thread of the argument, "the other four competitors must all perform on both of these days. With only the first day available, they cannot perform for one another. So at least 4 days are needed."

"A 4-day competition may be organized as follows," I said. "There are 16 different combinations of days. We print them on 16 tickets and give one to each competitor, who will perform on the days listed on the ticket. Clearly, no competitor can hold a ticket such that every day on the ticket is on the ticket held by another competitor. Such forbidden pairs differing in only one day are joined by line segments in the diagram below. We give each of the 6 competitors one of the tickets 12, 13, 23, 14, 24 and 34."

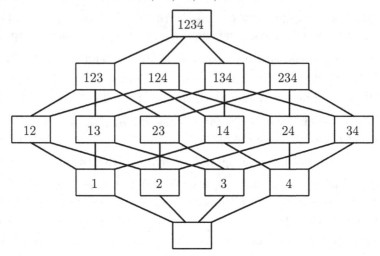

"Why is it impossible to accommodate 7 competitors in a four-day festival?"

"Every ticket belongs to at least one of the 6 overlapping chains in the diagram below, going from the empty ticket to the full ticket. If there are 7 competitors, 2 of them will be given tickets in the same chain. Then one of them will not be able to listen to the other one perform."

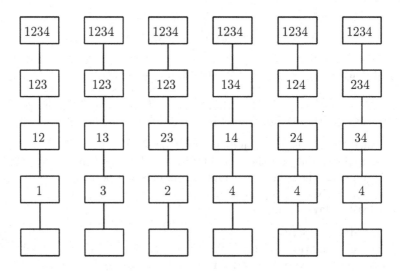

Chapter 18

The Banquet Tickets

Demeter and Persephone came over to the gatehouse for dinner that evening. Hebe prepared some roast beef, which was the favorite food of Heracles. Demeter, like the other Olympians, loved the smell but not the taste. This was why mortals sacrificed cattle to them, and afterwards ate the meat themselves.

Persephone, deprived while in Tartarus, enjoyed any kind of food above ground. Hebe liked roast beef too. I did not have much chance of eating that back home. Now I had an opportunity to feast on it. Hebe baked some barley cake for Demeter, and offered her some nectar.

"Persie," I asked her after dinner, "could you tell us more about Tartarus? Better still, draw us a map of the place."

"Here it is, Eubie."

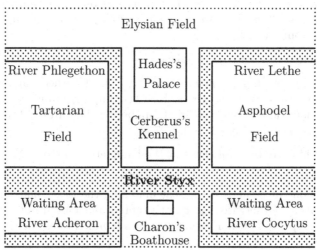

"All I knew is that Tartarus lies deep below the earth," said Hebe, "on the bank of the underground River Styx. From the map, I see that there are four streams which feed into the Styx. Tartarus is divided into three fields."

"When mortals die, Hermes leads their souls to the Western Ocean and down a dark tunnel. At the Styx, they pay Charon the boatman to row them across, using the coins which relations had placed beneath the tongues of their corpses. Those who cannot pay shiver in the waiting areas, until money is sent to them somehow. On the opposite bank of the Styx is the enormous three-headed dog Cerberus, who lets no ghosts escape and prevents any living mortal from entering."

"The two-headed dog Orthrus I killed during my tenth labor was the younger brother of Cerberus," said Heracles.

"On their first arrival in Tartarus," continued Persephone, "ghosts are tried by the three Judges of the Dead in an office at the front of the palace. They are sent to the Elysian Field if their lives have been good. If their lives have been bad, they are sent to the Tartarian Field where they are whipped by the Three Furies, horrid witches, and suffer other forms of punishment. Those whose lives have been neither very good nor very bad are sent to the Asphodel Field. That is where the 49 Danaïds are now."

'Tell us about someone who is in the Tartarian Field," I requested.

"Perhaps the most famous criminal is Tantalus of Lydia," said Persephone. "He had been invited to visit the Mount Olympus Palace, but he abused his privilege by stealing some ambrosia, and shared it among his mortal friends."

"That did not seem serious enough," Hebe remarked.

"He had done a lot worse," Demeter said. "Not knowing that his theft had been discovered, he wished to maintain good relations by inviting us for dinner at his palace. He issued a banquet ticket to each Olympian, with seat numbers from 1 to 12."

"Is that rather formal?" I asked.

"Yes, too much so," agreed Demeter. "As it turned out, others wanted to go too. They were the Fates, Hestia, Amphitrite and Semele. Tantalus printed more tickets, but instead of numbering them from 13 to 18, the extra tickets bore the numbers 2, 2, 3, 5, 7 and 11."

"Would there be total chaos at the banquet?" I asked.

"The seats were correctly numbered 1 to 18," said Demeter. "The guests sat down one at a time. If the seat matching the ticket number was vacant, it was taken. Otherwise, the guest grunted and moved onto the next seat. This was repeated until a vacant seat was found. It was then taken, without grunting."

"That still did not seem serious enough," remarked Hebe.

"Meat was served," said Demeter. "If Tantalus were testing whether we were omniscient, he was disappointed. We could tell at once that it was human flesh. Moreover, it was from the dead body of his murdered nephew Pelops!"

"That was more than serious," shouted Hebe. "It was an outrage!"

"Zeus killed Tantalus with a thunderbolt, and restored Pelops to life."

"Down in Tartarus," Persephone picked up the tale, "the judges sent Tantalus to the Tartarian Field to be whipped by the Furies every day, one lash for each time an Olympian had grunted during the seating at the banquet."

"I have heard of something called a Tantalus Punishment. Is this it?" asked Heracles.

"No," said Persephone. "Tantalus was tied to a fruit tree beside the Styx. Pears, apples and pomegranates dangled before him from branches, but whenever he tried to pick them, a wind always swept the branches away. Moreover, the waist-high river water always sank out of reach whenever he bent to drink. Thus he suffered from hunger, thirst and disappointment forever. This is the Tantalus Punishment."

"Serves him right!" exclaimed an indignant Hebe.

That brought the conversation to a close. The general mood was somewhat gloomy. I secretly regretted bringing up the subject.

Back at the gatehouse, Hebe and I turned our thoughts to something we both loved, solving puzzles.

"How many lashes did Tantalus get because of the grunts, Eubie?" she asked.

"I don't know off hand. The number may depend on the order in which the guests took their seats."

"In that case, what would be the maximum and minimum number of lashes?"

Solution

I first worked out an example.

Ticket Number	1	2	3	4	5	6	7	8	9
Number of Grunts	0	0	0	0	0	0	0	0	0
Ticket Number	10	11	12	2	2	3	5	7	11
Number of Grunts	0	0	0	11	12	12	11	10	7

•

The total number of grunts was 63. Hebe showed me her example.

Ticket Number	1	2	2	2	3	3	4	5	5
Number of Grunts	0	0	1	2	2	3	3	3	4
Ticket Number	6	7	7	8	9	10	11	11	12
Number of Grunts	4	4	5	5	5	5	5	6	6

The total number of grunts was also 63.

We stared at each other.

"Could it always be 63," I finally asked, "independent of the order in which the guests take their seats?"

We thought for a while. Finally Hebe said, "A grunt was made by a guest at a seat. Instead of calculating the number of grunts made by each guest, why don't we calculate the number of grunts made at each seat? For seat number 1, there was only one ticket whose number was at most 1, namely 1. Hence the number of grunts made at seat number 1 was $1 - 1 = 0$."

"I think you are onto something," I said in excitement. "While the number of grunts made by each guest clearly depended on the seating order, the number of grunts made at each seat clearly did not, but the two totals must be the same. For seat number 2, there were 4 tickets whose numbers were at most 2, namely, 1, 2, 2 and 2. Hence the number of grunts made at seat number 2 was $4 - 2 = 2$. For seat number 3, there were 6 tickets whose numbers were at most 3, namely, 1, 2, 2, 2, 3 and 3. Hence the number of grunts made at seat number 3 was $6 - 3 = 3$."

We drew up the following chart, which showed that the total number of grunts would always be 63.

Seat number	1	2	3	4	5	6	7	8	9
Number of Tickets	1	4	6	7	9	10	12	13	14
Number of Grunts	0	2	3	3	4	4	5	5	5
Seat Number	10	11	12	13	14	15	16	17	18
Number of Tickets	15	17	18	18	18	18	18	18	18
Number of Grunts	5	6	6	5	4	3	2	1	0

Chapter 19

The Eleventh and Twelfth Labors of Heracles

Heracles had the next day off. As I had worked out before, the moving on the Creamy Way came to an end. He took a well-deserved afternoon nap. When he woke up and found nothing waiting for him to do, he became listless. I took the opportunity and got him to tell me about his last two labors.

"Eubie, the first of the extra labor was to get three golden apples from the Garden of the Hesperides on the island where Geryon was, out in the Western Ocean. Ladon, an unsleeping dragon, guarded the apple tree by coiling around it."

"That shouldn't be much of a problem for you, sir," I said. "You have been a specialist on serpents and dragons since babyhood."

"The difficulty did not lie there," Heracles acknowledged. "I knew that the garden was on the island, but it was well-hidden, and I had no clue where it might be. I went around, looking for information. I met some nymphs who had heard of my earlier exploits before. I received from them a piece of advice, a piece of information and a gift."

"What advice could the nymphs give you?"

"They said that it had be prophesied that any mortal, which I was at the time, who took the apples would die instantly. They advised me to find an immortal to do the job for me."

"The piece of information must be the location of the garden," I guessed.

"Indirectly," said Heracles. "They did not know where exactly the garden was either. However, a being known as the Old Man of the Sea knew."

"What gift did you receive from the nymphs?"

"The gift was 8 identical 5 × 9 nets," said Heracles. "In each 10 of the 45 squares were filled with thick fur. It was something like the diagram below, with fur on the shaded ones. However, I am sure the fur squares were not in the correct positions, and I have forgotten where they should be. Anyway, these nets turned out to be crucial to my eventual success."

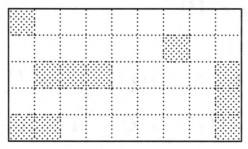

"How did you succeed?"

"I first caught the Old Man of the Sea. He tried to wriggle out of my grasp by transforming himself into various animals. However, I held tight, and he finally gave up and told me what I wanted to know. Then I found Atlas, the leader of the Titan army under old King Cronus during my father's revolt. The defeated Atlas was condemned to stand tall in the northwest corner of Africa, and carried the weight of the Heavens on his shoulders. He agreed to get the apples for me, on two conditions. First I had to kill Ladon. Second, to prevent the sky from falling while he went to get the apples, I had to stand in his place temporarily."

"The first condition was fine, but the second one might be a trick. He might just walk away and leave you there."

"That was his intention all along. Mostly likely, he would go and find other imprisoned titans, release them and start a revolt just as my father had. Anyway, I found the garden and from outside the wall, I shot an arrow and killed Ladon. On my way back to Atlas, I suddenly realized the significance of the nymphs' gift. I took a gamble and stood in for Atlas."

"Did he come back with the apples?"

"That he did," said Heracles. "He believed that I had fallen for his trick, and wanted to enjoy fooling me further. He said that with his titanic strides, he could bring the apples to Eurystheus faster than I could. He asked me to wait for his return."

"How did you wriggle out of this situation?"

"I told him to look inside my bag. He found the eight nets partially lined with fur. I told him that I wanted to stack them together into a 9×9 pad, filled with fur except for one corner square. I said I wanted to put the pad on my shoulders, to cushion them from the weight of the Heavens."

"Did he agree?"

"Yes," said Heracles, "He stood in for me and watched me assemble the pad. It was my turn to enjoy fooling him. I finished the task and showed him the assembled product before walking away."

'This is the most satisfying story among your labors," I said. "I hope the last one is even better."

"I am sorry to disappoint you. The twelfth labor was the easiest and most boring. I was to go to Tartarus and drag Cerberus to Eurystheus. Hades raised no objection provided that I used no weapons and did no permanent damage to his dog. Cerberus hated me for killing Orthrus, but it also feared me. It put up little resistance. When Eurystheus caught sight of its monstrous shape, he was scared stiff, and asked me to return the brute to Hades immediately."

Hebe called us back to the present, specifically, to the dining table. Afterwards, she and I tried to figure out what the eight nets looked like, and how they could be assembled into the furred pad.

"Save your energy," Heracles said to us. "I don't think I have thrown away those nets. It will take me a very short time to dig them out."

"Do that, my love," said Hebe, "but don't show them to us yet. We enjoy working things out. Later, we can check our results against your nets."

Solution

The problem was hard. There were too many ways of choosing 10 squares out of 45 to be filled with fur. Moreover, when the nets were superimposed on one another, the furred squares might overlap.

"One thing we do know," I said, "is that each net is contained within the 9×9 pad. Let me take 4 of them and stack them together as shown in the diagram below, covering all but the top row."

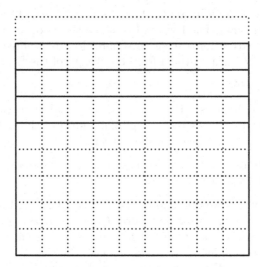

"I will take the other 4, Eubie, and stack them together in the direction perpendicular to yours, covering all but the leftmost column. Since the squares in the top row are not filled with fur from your nets, they must be filled with fur from my nets."

"Since the squares in the leftmost column are not filled with fur from your nets, they must be filled with fur from my nets. That leaves an 8 × 8 part in each of our pads."

"I have an idea," said Hebe after a while. "Divide the 8 × 8 part into 4 quadrants and filled the squares in 2 of them with fur."

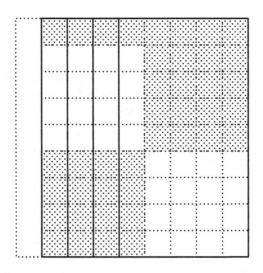

"You have done it!" I exclaimed. "I have drawn a new diagram. The number in each square indicates the net in which that square is filled with fur."

	8	7	6	5	8	7	6	5
1	1	1	1	1	8	7	6	5
2	2	2	2	2	8	7	6	5
3	3	3	3	3	8	7	6	5
4	4	4	4	4	8	7	6	5
1	8	7	6	5	1	1	1	1
2	8	7	6	5	2	2	2	2
3	8	7	6	5	3	3	3	3
4	8	7	6	5	4	4	4	4

As if on cue, Heracles brought the nets to us. They were exactly as we thought they would be.

Chapter 20

The Summer Festival

Heracles went back to work the next day, and found another funeral on his hands. This time, the Olympian who was an interested party was Artemis. He escorted her down to the Mount Olympus Cemetery, and I tagged along.

Artemis did not say a word on the way down. I observed her, and found that she was very different from her twin brother Apollo. Her demeanor was poised and, understandably in the current circumstances, subdued.

Heracles seemed to be in mourning as well. It turned out that the corpse belonged to his friend Orion, a fellow Theban.

Artemis gave a very brief eulogy.

"Orion was the handsomest man and the cleverest hunter. We met in Crete, and fell in love. One day, we agreed to go hunting together and arranged to meet by the sea. There was an accident, and I shot him. His body had only just been recovered. May you rest in peace, my love."

She broke down in tears.

During the reception, the three of us sat at a quiet corner. Although Artemis had seen me several times and knew who I was, we had not yet been formally introduced until now.

"Artemis, my dear," said Heracles, "there is something about the accident you haven't told us."

"My dearest Heracles," she said, "how perceptive you are. I was duped by my own twin brother."

"Why would Apollo interfere in your personal life?" asked Heracles.

"He was jealous of Orion," said Artemis. "He wanted to prevent me from marrying Orion, a mere mortal. Having found out about our hunting date, he sent a scorpion to attack Orion, who had arrived by sea ahead of me."

"Was he killed by the scorpion? I thought you said you shot him."

"I did. Unable to fight off the scorpion, Orion escaped by jumping into the sea and swam away. When I arrived, Apollo was there. He pointed to a round object far from shore, and lied to me that it was the head of a wretch called Candaon. He said that the creature had insulted a priestess in one of my temples. I shot an arrow through the head, ..." Artemis broke down again.

"I had met Orion during my visit back to Thebes," said Heracles. "He seemed to have bad luck all the time. He had told me that he once fell in love with Merope, daughter of Oenopion, King of Chios. Oenopion promised that he could marry Merope if he would kill all the wild beasts on the island."

"He never told me about that," said Artemis.

"That was because nothing came of it. Orion did his part, but Oenopion did not live up to the bargain. He sent someone to attack Orion by stealth, and poked out his eyes."

"When we met in Crete," said Artemis in surprise, "Orion was not blind."

"His friend Phaëthon was the son of Helius the Sun, whose palace was near Colchis beyond the Black Sea. Phaëthon took Orion there to have his sight restored by Helius. Orion went back to Chios for vengeance. Warned of his arrival, Oenopion hid in one of 91 farms on a circular road."

"It would take Orion a lot of time to search each of them," I said.

"Orion did not have the authority to search them. He could attack them, but that would turn the local populace against him."

"Could he just attack the one in which Oenopion was hiding?" asked Artemis.

"He didn't know in which farm Oenopion was hiding. As it turned out, there was a summer festival on this road. Each farm would bake an enormous barley cake. On each day of the festival, a farm with a cake would give it to the next village on either side, chosen independently. A farm receiving two cakes on the same day would feast on one of them. The festival would continue until only one cake remained, or by day 90 if there were still several cakes around. It was prophesied that the number of cakes would be down to 1 by day 90 that year, and Oenopion would be hiding in the farm with the last cake."

"Since several cakes could be feasted on in one day, Orion might not have to wait 90 days," I observed.

"That was what he expected too. However, when this did not happen by day 60, Orion lost faith in the prophecy and went on to Crete in search of his enemy."

"So that was how our paths crossed," remarked Artemis.

We consoled Artemis as much as we could. After the reception, Heracles tidied things up, and we made our way slowly back to the palace.

I went straight to the gatehouse and told Hebe the new problem.

"The artificial 90 limit was necessary," Hebe observed. "If there were 2 cakes left, and they kept moving in the same direction, they would never come together. The festival would not have ended."

"So there was no natural upper limit to the length of the festival. What would be a lower limit? In other words, what was the minimum number of days the festival could last?"

Solution

By now, we had established a routine. We downsized the number of farms from 91 to 9, and each of us worked out an example, trying to make the festival as short as possible. We numbered the farms 1 to 9, with farms 1 and 9 also neighboring.

In my example, the festival lasted 8 days.

Farms								
1	2	3	4	5	6	7	8	9
○→	○→	←○	←○	○→	○→	○→	←○	←○
	○→	○→			○→	←○	←○	
		○→	○→		←○	←○		
			○→	←○	←○			
			←○	○→				
		←○			○→			
	←○					○→		
←○							○→	
								○

This was also the case in Hebe's example.

Farms								
1	2	3	4	5	6	7	8	9
←○	←○	←○	←○	←○	○→	○→	○→	○→
←○	←○	←○	←○			○→	○→	○→
←○	←○	←○					○→	○→
←○	←○							○→
○→								←○
	○→						←○	
		○→				←○		
			○→		←○			
				○				

We stared at each other.

"Could the lower bound always be 8 days if there were only 9 farms," Hebe finally asked, "independent of how the cakes were passed around?"

I said after a while. "I think that based on your example, which is more systematic than mine, I can show that with 91 farms, the festival could have lasted only 90 days."

"How?"

"In the first stage, farms 1 to 46 passed their cakes through one another towards farm 91 while farms 47 to 91 passed their cakes through one another towards farm 1. This took 45 days, at which time there were only two cakes left, one in farm 1 and one in farm 91. In the second stage, these cakes were passed towards each other, ending with just one cake in farm 46 after another 45 days. Could this number be reduced?"

We thought long and hard. Finally, Hebe said, "This problem is really diabolical. Allowing the farms to eat the cakes is destroying evidence. Let us change the rules so that instead of eating any extra cakes received on a day, a farm must pass all received cakes to the same neighbor. The objective then is to have all the cakes in the same farm. The number of days required will be exactly the same."

"This is a wonderful idea," I said. "I think I can see how to continue. Consider what happens to a cake on two consecutive days. Either it returns to the farm which has it initially, or it is passed to a farm two places away. Suppose the cakes all end up in farm 46 in at most 90 days. There are not enough days for the cake initially in farm 46 to go once around the circular road and back. Hence the number of days in the festival must be even. However, in order for the cake initially in farm 45 to end up in farm 46 in an even number of days, it must go once around the circular road, and that takes 90 days!"

Chapter 21

The Great Amoeba Escape

"Oh, dear me, Eubie," Hebe lamented. "We have yet another funeral down in the Mount Olympus Cemetery."

"Who died this time?" I asked.

"There were actually two deaths, mother and son, but only one funeral," said Heracles. "Artemis took revenge on Apollo for the death of her lover Orion. She shot Coronis, a Thessalian woman whom he had married and by whom he had a boy."

From my impression of Artemis, I was surprised to hear that she could kill someone. Then I remembered that Athene had said that while Apollo was tame, he did have a mean streak. Perhaps Artemis, as Apollo's twin sister, had that in her as well.

"Did she kill the boy too, sir?" I asked.

"No, that was another story," said Heracles. "This funeral is not for Coronis. She was a mortal woman, and Apollo did not love her all that much anyway. It is for poor Asclepius. Let us get going, or we would be late."

The eulogy was given by Apollo. "My son Asclepius was educated in Mount Pelion under Cheiron, king of the centaurs. Asclepius was taught archery, the alphabet and astronomy. However, he was primarily interested in medicine, and grew up to become the best doctor in all of Greece. Not only did he cure dying people, but on several occasions, he was able to restore dead people to life. We gather here today to mourn his premature death."

At the reception, Apollo sat with Heracles and me at a quiet corner.

"Who would kill someone like Asclepius?" I asked no one in particular.

"Father struck him down with a thunderbolt!" exclaimed Apollo. "Uncle Hades complained that Asclepius had been stealing his subjects."

"That was a rather flimsy excuse for such a drastic action," I said.

"That was not the reason. Father did argue with Uncle Hades that Asclepius did those miracle cures out of the goodness of his heart. Moreover, all his patients would die eventually. The real culprit was Dionysus. Lycurgus, a Thracian general under King Midas, had defeated Dionysus's army as it came back in triumph from India. In revenge, Dionysus had Lycurgus torn to pieces by wild animals. Asclepius put him back together again, and the grateful Lycurgus paid him some gold."

"It is only reasonable for doctors to get paid," I remarked. "How else are they supposed to live?"

"Dionysus took some wine to father and drank with him. That did the trick. The old fool killed my son just to please Dionysus. I am not taking this lying down!"

In a blind rage, Apollo went and killed all the cyclopes who made the thunderbolts for Zeus. Zeus sentenced him to become for one year the slave of King Admetus of Pherae, a mere mortal.

Apollo considered this an insult and absolutely refused. Zeus suspended Apollo's seat on the Olympian Council and demoted him to a minor god. Soteria and Eleos were two minor goddesses who were vying to become the Goddess of Redemption, but Zeus had not made up his mind. In the meantime, Apollo could fill the vacancy as the temporary God of Redemption.

Apollo was in a foul mood. Heracles asked Eros and me to take him down to the Mount Olympus Arcade to take his mind off his troubles.

The arcade was on the opposite side of the foothills to the Mount Olympus Inn, which was why I had not known of it before. There were many stations inside, each featuring a game. Those who won were rewarded with wreaths which could be used to redeem prizes in the form of vases. The level of difficulty was indicated by the number of wreaths rewarded.

Most stations offered 5 to 10 wreaths, but I found one offering 100. It was called the "Great Amoeba Escape". The playing board was divided into unit squares. It was bounded by walls along the south and west edges, but theoretically extended infinitely to the north and to the east.

Initially, a single amoeba was on the square at the southwest corner. In each move, any amoeba could split into two, one moving one square to the north and the other moving one square to the east, provided that those squares were vacant. Success came when all 6 shaded squares in the diagram below had been vacated.

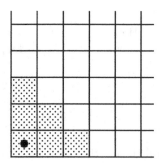

With only one amoeba initially and an infinite amount of vacant space, I wondered why this game should prove so difficult.

After a few futile attempts, I found that the board got clogged up very quickly. I began to suspect that this game could not have been won, but I could not prove that either.

Meanwhile, each of Apollo and Eros had collected many wreaths from much easier games. As we approached the redemption booth, there was a long line. Being a god, Apollo went straight to the counter, and learnt that one of the two clerks had been taken ill, causing a delay.

A stout man was next in line. He said to Apollo, "Whoever you may be, get in line like the rest of us!"

"I am the God of Redemption!" Apollo yelled.

"It is about time you show up," said the man, pushing Apollo behind the counter. "Start working. Here are my wreaths."

"But, but," Apollo stammered, taken aback. "That is not what the *God of Redemption* is all about."

"I don't care. Just give me that vase on the third shelf."

Overwhelmed by the situation, Apollo meekly handed the man what he wanted to redeem. It was only after serving three other clients when he noticed that I was still there. Eros had gone to the end of the line with his wreaths.

"This is an even bigger insult!" Apollo said to me at the top of his voice. "Eubie, you are now the temporary God of Redemption. If father comes to check on me, tell him that I am on my way to Pherae!"

"Me? A temporary god?" I gulped.

However, Apollo had gone. I did not like the look of the people in the line. So I went behind the counter, sat down beside the other clerk, and started redeeming prizes. Unlike Apollo, I did not particularly mind, and actually enjoyed this unexpected change of routine. By the time Eros got to the head of the line, very few people were behind him. After he had collected his vases, we left the arcade.

Back at the gatehouse, Hebe and I got a good laugh about the temporary God of Redemption. Then we started working on the "Great Amoeba Escape".

Solution

"At the start, Eubie," said Hebe, "there was only 1 amoeba. After one move, there were two. Their number kept increasing. However, it would be easier to keep track of things if each new amoeba was considered to be $\frac{1}{2}$ of the original one. This way, there would always be 1 amoeba. Similarly, a $\frac{1}{2}$-amoeba became two $\frac{1}{4}$-amoebae, and so on."

I said, "Clearly, how much a piece of the amoeba depended on its location. We should assign values to the squares themselves, as shown in the diagram below."

$\frac{1}{16}$	$\frac{1}{32}$	$\frac{1}{64}$	$\frac{1}{128}$	$\frac{1}{256}$
$\frac{1}{8}$	$\frac{1}{16}$	$\frac{1}{32}$	$\frac{1}{64}$	$\frac{1}{128}$
$\frac{1}{4}$	$\frac{1}{8}$	$\frac{1}{16}$	$\frac{1}{32}$	$\frac{1}{64}$
$\frac{1}{2}$	$\frac{1}{4}$	$\frac{1}{8}$	$\frac{1}{16}$	$\frac{1}{32}$
1	$\frac{1}{2}$	$\frac{1}{4}$	$\frac{1}{8}$	$\frac{1}{16}$

"Call the total value $1 + \frac{1}{2} + \frac{1}{4} + \frac{1}{8} + \cdots$ of the squares in the first row the magic number. Then twice this magic number is equal to $2 + 1 + \frac{1}{2} + \frac{1}{4} + \frac{1}{8} + \cdots$, which is 2 more than the magic number itself. Aha!" exclaimed Hebe, "The magic number is equal to 2."

Since each square in the second row was half in value of the corresponding square in the first row, the total value of the squares in the second row was 1. Similarly, the total values of the squares in the remaining rows were $\frac{1}{2}$, $\frac{1}{4}$, $\frac{1}{8}$, ... Hence the total value of the squares on the infinite playing board was only 4.

The total value of the 6 shaded squares was $2\frac{3}{4}$. To win the game, all the partial amoebae had to fit into the unshaded squares. Their total value was $1\frac{1}{4}$, so that there was not a lot of room to play about.

At any time, each of the first row and the first column had exactly one partial amoeba. If it was not on a shaded square, its value was at most $\frac{1}{8}$.

The remaining squares with total value $\frac{1}{16} + \frac{1}{32} + \cdots = \frac{1}{8}$ were wasted. Hence the total value of the unshaded space not in the first row or column was only $1\frac{1}{4} - 2(\frac{8}{+}\frac{1}{8}) = \frac{3}{4}$. However, there was still $1 - 2 \times \frac{1}{8} = \frac{3}{4}$ of an amoeba to be accommodated. Thus game could not be won.

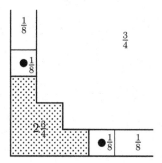

Chapter 22

The Rolling Stones

That evening, the three of us went over for dinner at Demeter's place. It was a big gathering. Hephaestus, Ares, Athene and Hermes were there. Apollo was in Pherae and Dionysus had not been invited. Artemis and Aphrodite sent their regrets.

Afterwards, I asked Persephone for another story about Tartarus.

"Please, not a gruesome tale this time," said Hebe. "Eubie has a very sensitive nature, you know."

"I can handle it," I said.

"This story will be to Eubie's liking," said Persephone. "From time to time, the people in the Asphodel Field were allowed to visit the Elysian Field. They might also be taken to the Tartarian Field for some mild form of punishment. Sisyphus, the King of Corinth, was in the Asphodel Field. He had actually done a lot for his people. He built the Corinthians their first fleet. They loved him, and held an annual feast in his honor."

Hermes said, "I met him when father sent me to Corinth to settle a dispute Sisyphus had with his neighbor Autolycus. Autolycus had been kind to my mother before I was born. He let her hide in his house in order to evade the jealous Hera who tried to kill her. In gratitude, I gave him the magic powers of turning bulls into cows, or vice versa, and changing their colors, say from white to red, or from black to piebald. You all know that I had had some trouble with Apollo and his white cows when I was a baby. I only learnt these new magic powers since then."

"What was the dispute?" asked Ares, who was always interested in disputes.

"Sisyphus owned a large herd of cattle while Autolycus owned a smaller one. Autolycus stole some from Sisyphus and used these magic powers to cover up his crime. Sisyphus watched his herd getting smaller, and Autolycus's getting bigger day-by-day. He brought charges against Autolycus."

"Could he prove it?" asked Athene.

"Sisyphus marked the hooves of his remaining cattle with the letters $\Sigma I \Sigma$, the first three letters of his name. Five cows so marked were found in Autolycus's herd."

"That should settle the matter," said Athene.

"Not so easily," said Hermes. "Autolycus claimed that Sisyphus had come into his pastures and marked the hooves of his own cows. Everyone argued and shouted. Meanwhile Sisyphus had his revenge. He slipped into Autolycus's house and seduced one of Autolycus's daughters."

"Was that why he ended up in the Asphodel Field?" I asked.

"No," said Hermes. "Autolycus did not love that particular daughter. He considered it a good bargain. I told him to stop stealing cattle from Sisyphus. The trouble came from Asopus, a river god. He had just lost a daughter, and accused Sisyphus of having seduced her as well."

"I guess losing daughters is about as common as losing cattle," observed Demeter.

"Sisyphus had not seduced Asopus's daughter," said Hermes, "but knew where she was. He said he would tell Asopus where she was if Asopus would make a spring break out from the hill on which Corinth stood. Asopus struck the ground with a magic club and made a spring appear. Sisyphus then told him that someone had fallen in love with his daughter, and that they were walking arm-in-arm in a wooded valley nearby."

"Who could that be?" asked Hephaestus. "Not father in his philandering ways again?"

"Who else?" said Hermes. "Asopus, very angry, went in search of father, who had left his thunderbolts hanging on a tree. So father disguised himself as a round stone. After Asopus had rushed past, father turned back into his true shape, fetched his thunderbolt and threw it at Asopus, wounding him in the leg. Father then ordered Uncle Hades to arrest Sisyphus and put him in the Tartarian Field. The charge was that he had betrayed a divine secret to Asopus."

"That was unfair," said Hebe. "Father was both the prosecutor and the judge."

"Uncle Hades came to see me before he went to Sisyphus," said Hephaestus. "I had just invented handcuffs by chaining steel bracelets together. Uncle

Hades was afraid that Sisyphus might refuse to follow him, and thought that my handcuffs would come in handy."

"Did they help?" I asked.

"No," said Ares. "I soon found that battles had become sham fights because nobody got killed. I went to Corinth to investigate. At Sisyphus's place, I found Uncle Hades handcuffed and chained up in a dog's kennel. He said that Sisyphus had tricked him, asking him to show how handcuffs worked. Stupidly, he put them on himself, and Sisyphus quickly locked them. Then he unchained his dog and fastened the dog's collar around Uncle Hades's neck. I threatened to strangle Sisyphus. He pointed out that he could not be killed as long as Hades was chained up. I told him that I did not intend to kill him, but squeeze his throat until his face would turn black and his tongue would stick out. Sisyphus had no choice but to obey, but he still refused to follow Uncle Hades, objecting that Hermes was the god who fetched ghosts."

"I returned to Corinth again," said Hermes, "and confirmed that the Fates had indeed cut off Sisyphus's thread of life. He sighed and followed me. He said that at any rate, he got a fine spring of water for Corinth."

"The Judges of the Dead decided that Sisyphus should be sent to the Asphodel Field," said Persephone.

"Was my brother satisfied with that?" asked Demeter.

"No, but he could not change the judges' decision," replied Persephone. "As I have mentioned earlier, it was possible for Sisyphus to be sent to the Tartarian Field for a period of time to receive some mild form of punishment. Zeus had Hades prepare a hill with 1001 steps. There was a round stone on each of the first 500 steps, in exactly the same shape as the one which Zeus turned into during his escape from Asopus. In each move, Sisyphus might push any stone to the lowest vacant step above it. After each move, one of the stones would roll down to the step immediately below, provided that it was vacant. Sisyphus could return to the Asphodel Field if he could push a stone to the top step."

"Did Sisyphus succeed?" I asked. "He seemed to be full of tricks."

"He is still in the Tartarian Field," said Persephone. "I am sure Zeus had given him an impossible task."

Solution

When we went back to the gatehouse, Hebe and I worked on the problem.

"Let us start with a small hill with 11 steps and a stone on each of the first 5 steps. I will play Sisyphus and you, Eubie, plays the rolling stones.

In my first move, I can push any of the stones up to step 6. Obviously, my best choice is to push the bottom one."

"I don't have any choice in my first move. I must roll the stone on step 2 down."

"In my second move," said Hebe, "there is no point in my pushing up the stone on step 1. It can only go up to step 2, and you will roll it down again. As before, my best move is to push the stone on step 3 up to the step 7."

"Again, I don't have any choices. I must roll the stone on step 4 down."

"In my third move, I should not push the stones on either step 1 or 3, because it will come right back again. I should push the stone on step 5 up to step 8."

"This time," I said, "I have a choice. I can roll the stone on either step 3 or 6 down. I think that the latter is the better choice."

We made one more move, and recorded the progress of our game in the diagram below.

Steps	Moves 1		2		3		4	
9							●	●
8					●	●	●	○↓
7			●	●	●	●	○	●
6	●	●	●	●	●	○↓		
5	●	●	●	●	○	●	●	●
4	●	●	●	○↓				
3	●	●	○	●	●	●	●	●
2	●	○↓						
1	○	●	●	●	●	●	●	●

"I have pushed a stone as high up as step 9 so far. Now there is a stone on every other step below the top stone. I can push it up to step 10, but that is all, because you can cancel every move I make from now on. It would appear that Sisyphus was doomed."

However, what we thought were the best moves were not necessarily so. We needed a solid proof.

"In the original problem with 1001 steps," I said, "could I, controlling the rolling stones, be assured that after any move of mine, the first step would be occupied and no two consecutive steps below the top stone were vacant?"

"Well, the first step was occupied initially. If Sisyphus pushed the stone on it up, to step 2 or above, the step 2 had to be occupied. You could refill step 1. In order for Sisyphus to create two consecutive vacant steps, the lower one must already exist. However, as soon as he created the upper one, you could refill it."

"Then Sisyphus was indeed doomed. Whenever it was his turn, the top stone could be no higher than on step 999, so that he could never push a stone to step 1001."

Chapter 23

The Conscientious Birds

Time flew while I was enjoying myself. My month in Mount Olympus was coming to an end. I thanked Demeter and Persephone for inviting me, my hero Heracles for taking me under his roof, and Hebe, with whom I had shared many joyful moments in problem solving. I went around and bade farewell to the gods and goddesses, especially Eros, who had played with me on many occasions.

Silenus was waiting for me in the Mount Olympus Inn. He had arrived the night before, bringing with him a friend and a coffin.

"We meet again, Bias," I said to the friend. "Where is Pero? Don't tell me that she was inside that wooden box."

"Bias and Pero are fine, Eubie," said Silenus, "as far as we know. This is Melampus, Bias's twin brother. We met in Phrygia while I was visiting my former employer, King Midas. It is the king's body that is inside the coffin. He had left word that he wished to be buried in the Mount Olympus Cemetery."

"How did he die, sir?"

"Suicide, after he had murdered his barber. There was a third death at about the same time. Remember that Apollo barred a Phrygian shepherd named Marsyas from entering the Phrygian Music Festival. Marsyas issued a challenge to Apollo in a one-on-one music competition."

"Would Apollo accept the challenge from whom he called a riff-raff?"

"Apollo ordered Midas and the nine Muses to act as judges. Marsyas played the flute and Apollo the lyre. The judges could not at first agree which had given the better performance. Then Apollo challenged Marsyas to another competition, each playing his instrument upside down. Since the lyre was essentially the same upside down while the flute was definitely not, Apollo won nine-to-one. The only dissenting vote came from Midas, who said this was unfair."

"It most certainly was," I said.

"Apollo shot Marsyas through the heart for daring to challenge the God of Music himself! He flayed him, and gave his skin to me for making drums. Apollo then called Midas an ass, and touched his ears, which became long and hairy like that of an ass."

"That must have been embarrassing, especially for a king."

"To make things worse, the Phrygians cut their hair very short. In order to keep his secret, Midas had to wear a tall Phrygian cap all the time."

"His barber must know. You said earlier that he had murdered his barber. Did he let out the secret?"

"I can explain," said Melampus, taking over the conversation from Silenus. "The barber, bursting with the secret, dug a hole in the bank of the Pactolus River, looked carefully to be sure that nobody was around, and then whispered into the hole about Midas's ears. He filled up the hole at once to bury the secret. However, a reed sprouted from the hole, and whispered to 64 birds, one at a time."

"The birds told you the secret then?"

"Not right away," said Melampus. "The birds were conscientious. The head-bird, not among the 64 herself, wanted every bird to be well-informed before reporting the news. A bird was well-informed when she knew that every other bird had heard the news. The head-bird ordered the others to meet in pairs, and continue meeting until all of them were well-informed."

"Can you explain with an example?" I asked.

"Suppose that there are only 4 birds. Call them A, B, C and D. A must be well-informed, meaning that she knows that each of the others, namely B, C and D, has heard the news. The same applies to each of B, C and D as well."

"With 64 birds, wouldn't that take 63 days?"

"No," said Melampus. "It took a lot less than that, because when two birds met, they shared the list of birds either of them knew to have heard the news. To continue with my example, if A knows that B has heard the news and C knows that D has heard the news, then when A and C meet, A will know about D from C, and C will know about B from A."

"I see. So when all 64 birds became well-informed, they reported to the head-bird, and she told you the whole thing. Did you tell anyone?" I asked.

"I had been indiscreet," admitted Melampus. "I casually mentioned it to a few friends, and word spread. One day, Midas was driving in his chariot outside his palace when he heard his people shouting in chorus. They asked him to remove his cap and show them his ears. He returned to his palace, cut off the barber's head and then stabbed himself in the heart."

Athene, Apollo, Hermes and Dionysus came down to attend the funeral. Silenus and I sat in the lobby, and we worked out the smallest number of days required by 64 birds to confirm the news they had heard.

Solution

I arranged Melampus's 4 birds in a 2×2 table. Each square represented one of them, and its content was the the list of birds known to this particular bird to have heard the news. Silenus called this her knowledge base.

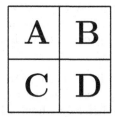 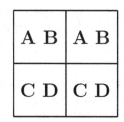

A B	A B
C D	C D
A B	A B
C D	C D

"Before any meeting takes place, Eubie," he said, "each bird knows only that she herself has heard the news. The knowledge bases are as shown in the diagram above on the left."

"On the first day, sir," I said, "the two birds in the same row meet. After the exchange of information, their knowledge bases have increased to that shown in the diagram above in the middle. On the second day, the two birds in the same column meet, and the new knowledge bases are shown in the diagram above on the right. Now the confirmation process is complete, in just two days."

"Excellent, Erbie." said Silenus. "How many days would the confirmation process take for 64 birds?"

"Initially, each knowledge base contains the name of only one bird. After each day, its size can at most double. It follows that for 4 birds, 2 days are necessary. For 64 birds, 6 days would be necessary. Would 6 days be sufficient?"

"Certainly enough. The birds could have been arranged in an 8×8 table. On day 1, information was exchanged between the birds in columns 1 and 2, between those in columns 3 and 4, between those in columns 5 and 6, as well as columns 7 and 8. On day 2, information was exchanged between the birds in columns 1 and 3, between those in columns 2 and 4, between columns 5 and 7, as well as columns 6 and 8. On day 3, information was exchanged between the birds in the columns 1 and 5, between those in columns 2 and 6, between columns 3 and 7, as well as columns 4 and 8. Then every bird would know that the other 7 birds in the same row had acknowledged that they had heard the news."

"I see now. In the last three days, we repeat the above steps, with rows and columns interchanged. Thus the confirmation process could be completed in 6 days."

"What would happen if the head-bird had also heard the news from the reed, and had to be included in the confirmation process?"

"Off hand, I would say it would take an extra day."

"No," said Silenus, "7 days would not be enough. You had pointed out earlier that the size of a knowledge base could at most double in a day. With 65 birds, no knowledge base was complete after 6 days. On day 7, since the number of birds was odd, at least one of them did not take part in any meeting on the day. Her knowledge base could not be complete until day 8 at the earliest."

"And 8 days are enough! On day 1, the head-bird met one of the others who would include the head-bird in her knowledge base. During the next 6 days, the confirmation process was carried out as before, without the head-bird. After day 7, the knowledge base of every bird was complete except for the head-bird. On the final day, one of the other birds shared her complete knowledge base with the head-bird."

Chapter 24

The Archipelago Tour

Silenus checked out after lunch. Demeter and Persephone met us in the lobby. I thought they had come to bid us farewell. However, they had packed their bags too, and were leaving with us.

"Too much is happening here," complained Demeter. "In the past month, we had *four weddings and a funeral.*"

"You mean a wedding and four funerals, mother."

"You see, my mind has been affected already. I want a complete change of scenery, away from Mount Olympus. We are taking a vacation."

Demeter's chariot was ready to take off, and she offered to give us a ride to Eleusis. We accepted eagerly. Then I decided, on the spur of the moment, to join them on their vacation as well. Silenus was happy to extend his own vacation. Demeter and Persephone were delighted.

On our way south, we landed at Pherae to check on Apollo. We found him in a good mood.

"Are you being well treated here, nephew?" Demeter asked.

"Better than an honored guest, auntie," replied Apollo. "King Admetus and Queen Alcestis are old friends of mine. Alcestis was the youngest and the prettiest of the daughters of King Pelias of Iolcus. There were many suitors. Pelias announced that he would give her in marriage to anyone who could harness a wild boar and a lion to a chariot and drive it around. Many tried and failed. Admetus prayed to me. I brought my lyre and charmed the wild boar and the lion with music. Admetus had no trouble harnessing them and driving the chariot around."

Admetus was only too willing to give Apollo permission to join us.

By the most amazing coincidence, my brother's mission took him through Pherae. We had a most delightful reunion. Triptie was on the verge of puberty when he left home. He had grown a lot since then, being taller than Persephone already. His old rambunctiousness had been replaced by a rather unexpected reserve. I felt a little awkward too, as I had not changed much, being very much a child still. Demeter told Triptie that he could have some time off and join us.

We went out to the Sporades Archipelago not far away. Tourism had become an increasingly important business here, and the tourists' map showed 109 islands with tour boat service. These boats connected various pairs of islands, running both ways, and from any island, it was possible to travel to any other island in one or more boat rides.

We decided that we would not visit any island more than once. Demeter would choose the island where we would start. After that, Apollo and she would choose alternately the next island we would visit, Apollo getting the first choice. Of course, it had to be accessible by boat, and had not yet been visited by us.

At some point, we would arrive at an island which was connected by boats only to islands we had already visited. Demeter and Apollo agreed that whoever was to choose next would pay the bill from the tourism office.

We landed on one of the islands at dusk, and pitched tents for the night. We had three of different sizes. Apollo took the small one for himself. Silenus, Triptie and I crowded into the large one. Demeter and Persephone shared the remaining tent.

"I want to help Demeter, sir," I said to Silenus. "She is kind enough to invite us on this vacation. I don't want to cost her extra money. Apollo can afford it because he can put the charges on King Admetus's account."

"We'll try, Eubie," Silenus agreed, "but first we have to study the map."

Triptie said, "It may be too late. Demeter has already chosen the starting island, and this may be a bad choice."

"It is also possible that every starting island is a bad choice," said Silenus.

"Yes," said Triptie. "Suppose there are only 2 islands connected to each other. No matter where we start, Apollo can choose the other island for our next stop, and watch Demeter trying to find her purse."

"The advantage goes to Apollo because 2 is an even number," said Silenus. "Since 109 is odd, the advantage may come to us."

"Let us try a small example with 9 islands," I suggested. "I will number them 1 to 9, and let the boat routes be as shown in the diagram below."

"Once we have landed," said Triptie, "Apollo can choose whether to move left or right, unless we land on island 1 or 9. Thereafter, only one choice is available. If we land on an odd-numbered island, we win. Otherwise, Apollo wins. This is too easy. Try my example."

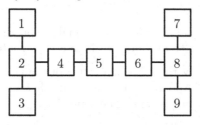

"If we land on island 1 or 3," I said, "Apollo must choose island 2 and we win in the next move. The situation is analogous if we land on island 7 or 9."

"We also win if we land on island 4 or 6," said Silenus. "Sooner or later, Apollo must choose island 2 or 8. However, if we land on island 2 or 8, Apollo wins on his next move."

"Finally," said Triptie, "if we land on island 5, Apollo can force us to choose island 2 or 8, and he wins on his next move. From these two examples, where we land does matter."

"We can't try every possibility when there are 109 islands," I said. "We have to find a better way."

Solution

We thought about it for quite a while. Then Silenus came up with a brilliant idea.

"Let us prepare what I call escape routes. Whenever Apollo's choice takes us to one end of an escape route, we follow it to escape to the other end of it. There is at most one escape route from each island, and we should prepare as many of them as possible. Note that escape routes go both ways, pairing two islands. Then we land on an island without an escape route."

"Can you make this clearer?" asked Triptie.

"Look at Eubie's example. We can prepare four escape routes, marked by double lines in the diagram below."

"Since 9 is odd," I said. "there is always at least one island not on an escape route. The only one here is island 5. Every choice by Apollo leads us to an island with an escape route, which allows us to escape. From there, Apollo will lead us to another island with an escape route, and so on. This works!"

"Not so fast," said Triptie. "What makes you think that Apollo's choice will always lead you to an island with an escape route?"

"This is certainly the case right after our landing, on an island without an escape route. If Apollo can lead us to another such island, we can add the route connecting them as an escape route."

"What about later?" asked Silenus.

"Consider my example, where two escape routes are marked by double lines in the diagram below. No more escape route can be added, or some island will have two."

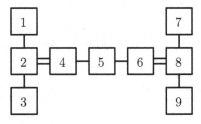

"Suppose we land on island 5," I said. "Apollo leads us to island 4 and we escape to island 2. Then he leads us to island 1 and wins! If we land on island 1, Apollo leads us to island 2 and we escape to island 4. Then he leads us to island 5 and there are no escape routes there!"

That was indeed a stumbling block. We worked on it for a while. Then I decided to focus on the second failed tour in Triptie's example.

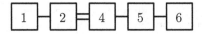

"This tour fails because you have not prepared the largest possible number of escape routes," I said.

"How can that be? You cannot add any escape routes to the ones I have prepared."

"That does not mean that there are no other ways," said Silenus. "In the second failed tour, if we replace the escape route between islands 2 and

4 by two other escape routes, between islands 1 and 2 and between islands 4 and 5, all conditions are still satisfied. Now the number of escape routes has increased by one. The diagram below shows how your example would look like. If further increases are possible, which is not the case here, we carry them out until we have prepared the most escape routes possible."

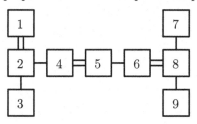

"Now we can land on island 3, 7 or 9. Following our strategy, we will win by escaping to island 1, sooner or later."

We took out the actual tourists' map, on which the tour boat routes were clearly marked. We prepared the largest possible number of escape routes. We were delighted to discover that we had started on an island without an escape route. Demeter's mind might have been affected, but apparently not impaired.

Chapter 25

The Fireballs and Lightning Bolts

My parents were very glad to see me after my month-long absence. Day after day, they asked about all my adventures. They were thrilled to learn of my chance meeting with Triptie.

Mother and the royal cows had grown so fond of each other that they were inseparable. Thus I could continue my studies with Silenus without interruption. After my experience in Mount Olympus, he felt that I was ready for all the important classical literature.

One day, during a rest break, I asked, "What have you been up to, sir, after we parted at the Mount Olympus Inn?"

"As I told you, I looked up some old friends in the area, all satyrs. We drank wine and played drums together. After a while, I got tired of that, and returned to the south. When I passed through Eleusis, I told your parents that you were well taken care of by Heracles and Hebe. Then I went further south and visited King Perseus of Tiryns. His grandfather was Acrisius, King of Argos."

"Why didn't Perseus inherit his grandfather's throne?"

"Acrisius had been warned by an oracle that his grandson would kill him. He decided to have no grandchildren, and locked Danae, his only daughter, in a tower. Zeus fell in love with and married her, who gave birth to a son, and that was Perseus. Acrisius dared not kill mother and child himself. He locked them in a wooden chest and threw it into the sea."

"Perseus must have survived, but how?"

"A fisherman on the island of Seriphos caught the chest and took Danae and Perseus to Polydectus, King of Seriphos. Polydectus, falling in love with Danae at first sight, let them stay, but Danae warned him to leave them alone or incur Zeus's wrath. However, Polydectus continued to try making Danae change her mind. When Perseus was 15 years old, he offered to get Polydectus anything he would want if Polydectus would stop pestering Danae. Polydectus asked for Medusa's head, thinking that he could get rid of Perseus."

"Who or what was Medusa?"

"She was a gorgon — a winged monster with 30 snakes for hair. Whoever looked at her would turn to stone."

"How could Perseus fight her if he could not even look at her?"

"Perseus prayed to Athene, who advised him to polish his shield until it shone like a mirror."

"Still, he would have to learn how to fight an enemy who was behind his back, without turning around."

"True," acknowledged Silenus. "Athene also gave him a bag which contained 100 fireballs in three different colors inside individual black pouches, and 111 lightning bolts in four different colors inside individual white pouches. She told him that he needed to throw thirty fireballs of the same color to singe Medusa's snake hair, and then three lightning bolts of different colors to kill her. The bag would be useful for holding Medusa's head after Perseus cut it off."

"How many of each color were there for the fireballs and the lightning bolts?"

"Perseus did not know. The color of an item could not be determined until it had been taken out of its pouch, but then it had to be used immediately. The bag came with a guarantee. Any 26 black pouches would contain at least 10 fireballs of the same color, and any 100 white pouches would contain at least 4 lightning bolts of different colors."

"That was most interesting," I said.

"Perseus decided to practice throwing fireballs and lightning bolts backwards. He calculated how many he could afford to use for this purpose, and had enough left to confront Medusa. When he did, he opened the black pouches and found 30 fireballs of the same color. After Medusa's snake hair had been singed, he opened the white pouches and found 3 lightning bolts of different colors. After Medusa had been killed, he cut off her head and put it inside the bag."

"How many fireballs and how many lightning bolts did Perseus use for practice?"

"Not too many, and you can figure out later the maximum numbers he could have used. Flush with his victory, Perseus lost his sense of direction, and went west instead of east. He did not realize his mistake until he came upon the titan Atlas, who threatened to drop the Heavens on him unless Perseus could persuade his father Zeus to release Atlas. Perseus showed Medusa's head to Atlas, who turned to stone and became the great Mount Atlas."

"Quite an appendix to Perseus's adventures," I remarked.

"He had another one. Turning back east, Perseus passed among the Philistines in Palestine, and rescued Princess Andromeda from a sea serpent. He married her and took her with him back to Seriphos. He discovered that Polydectus was still pestering Danae. Perseus turned him and his whole family to stone. Then he returned the bag with the head of Medusa to Athene. He himself went to Tiryns and became the king. He also built the famous city of Mycenae nearby."

"So he is living happily ever after then."

"There was only one sad note. He and his mother returned to Argos for a visit, and forgave King Acrisius. However, at an athletic competition, a sudden wind caught a quoit Perseus had thrown and blew it crashing through Acrisius's skull."

"Was Perseus punished for that?"

"No, it was after all an accident."

I turned my attention to figuring out how many fireballs and lightning bolts Perseus could afford to use for practice.

Solution

The information about the contents of the bag was given in a most unusual way. I needed to understand it properly. I called the colors of the fireballs red, yellow and blue. By symmetry, I might assume that there were at least as many red fireballs as yellow ones, and at least as many yellow fireballs as blue ones.

I knew that among any 26 fireballs, there would always be 10 of the same color. What could I learn about the largest possible number of blue fireballs? Suppose it was 8. We might have at least 9 yellow fireballs. Then 26 fireballs might consist of 9 red, 9 yellow and 8 blue, and there would not be 10 of the same color. Hence the number of blue fireballs was at most 7.

It was not enough for Perseus to save 65 fireballs to confront Medusa. There might be 47 red, 46 yellow and 7 blue fireballs in the bag. Among any 26 fireballs, at least 19 were red or yellow, so that there would be 10 of the same color. Yet, the 65 fireballs might consists of 29 red, 29 yellow and 7 blue, and there would not be 30 of the same color.

On the other hand, it was enough for Perseus to save 66 fireballs. Since there were at most 7 blue fireballs, there would be at least 59 fireballs that were either red or yellow, and 30 of them would be of the same color. Hence Perseus could use as many as $100 - 66 = 34$ fireballs for practice.

I now applied a similar argument to the lightning bolts. I called the colors red, yellow, blue and green, with at least as many red ones as yellow ones, at least as many yellow ones as blue ones, and at least as many blue ones as green ones.

I knew that among 100 lightning bolts, there would always be 4 of different colors. This time, I wanted to find the smallest number of green lightning bolts. Suppose it was 11. Then there were 100 non-green lightning bolts, which was contrary to the given condition. Hence the number of green lightning bolts was at least 12.

It was not enough for Perseus to save 87 lightning bolts to confront Medusa. There might be 75 red ones and 12 of each of yellow, blue and green ones in the bag. The total number in any three colors was at most 99, so that among 100 lightning bolts, there would be 4 of different colors. Yet, the 87 lightning bolts might consists of 75 red and 12 yellow, and there would not be 3 of different colors.

On the other hand, it was enough for Perseus to save 88 lightning bolts. Since there were at least 12 green lightning bolts, there would be at least 12 blue ones. Hence the total number of lightning bolts of any two colors was at most 87, and Perseus would have 3 lightning bolts of different colors. He could us as many as $100 - 88 = 12$ lightning bolts for practice.

Chapter 26

The Leaded Spears

"During your recent trip, sir," I asked, "did you meet any former heroes other than Perseus?"

"Quite a few, Eubie, and I did not even have to leave Tiryns. Being a former hero who had done very well, Perseus set up a sort of a retirement home for those who had fallen on hard times. The most notable one was Bellerophon of Corinth. He told me quite a story about his life."

"Tell it to me," I asked eagerly.

"Bellerophon was engaged to Princess Aethra. Like Perseus, he accidentally killed a man in a dart-throwing competition. He had to leave Corinth and fled to Tiryns. That was long before Perseus became its king."

"What a coincidence!" I exclaimed.

"The king at the time welcomed him as his guest. The queen fell in love with Bellerophon and asked him to elope with her. He wanted no part in an affair with a married woman. Moreover, her husband was the king who had been very kind to him. Afraid that Bellerophon might tell the king, she went to him first and accused him of trying to seduce her."

"What a spiteful woman!"

"The king believed the story, but dared not offend the Furies by killing his guest. So he asked Bellerophon to deliver a letter to the queen's father Iobates, King of Lycia. The letter asked Iobates to behead the messenger for having insulted his daughter."

"What did Iobates do?"

"He dared not offend Hermes, the God of Messengers. Moreover, it was an unsubstantiated charge. He decided to ask Bellerophon to perform some difficult and dangerous tasks for him. If he died in the process, then he would have satisfied his daughter and son-in-law."

"Since he is still living in Perseus's place, he survived."

"There was no task too difficult for him, even though killing the Chimaera was no ordinary task."

"Who or what was the Chimaera?"

"The Chimaera was a fire-breathing goat which guarded the palace of Iobates's enemy, the King of Caria. It had 7 lion's heads, and 4 more might appear now and then. Bellerophon needed to approach the Chimaera from the air. Since he hailed from Corinth, he knew of a wild winged horse called Pegasus, living by the spring which Sisyphus had asked the river god Asopus to create for the Corinthians. He decided to capture it."

"Wouldn't that mean that he had to return to Corinth, where he was still wanted for murder?"

"He revisited Corinth in secret. Waiting all night by the spring, he caught Pegasus coming for a drink, and tamed it after a fierce struggle. At that moment, his enemies came up to arrest him, but he mounted Pegasus and flew away."

"Was Pegasus enough for him to kill the Chimaera?"

"No," replied Silenus. "From above the palace in Caria, Bellerophon shot the Chimaera full of arrows, but range weapons had no effect on it. So he had to engage in melee combat, a daunting task. Bellerophon prayed to Athene. She told him to prepare a lump of lead. If he could get the lump into the Chimaera, its fiery breath would melt the lead. That would burn holes in its stomach and it would die."

"If he put the lump of lead at the tip of a spear and threw it into one of her open jaws, that should have done it."

"The Chimaera had more than one head. Athene told Bellerophon that an equal amount of lead must go down each jaw."

"But Bellerophon might be facing a 7-headed monster or an 11-headed monster," I said.

"Now you appreciate the difficulty of the task," said Silenus.

"He must divide the lump of lead into 77 units, and put one unit on the point of each of 77 spears."

"Bellerophon would never have time to use all of them. What actually happened was that he put an unequal number of units on each of several spears, and killed the Chimaera. Then Iobates showed him the letter from Tiryns, and stated that his elder daughter was always a liar. He married Bellerophon to his younger daughter, who was well-behaved as well as beautiful. Since he had no sons, he left the throne of Lycia to Bellerophon in his will."

"That seems like a happy ending," I said. "Why isn't Bellerophon living it up in Lycia rather than sulking in a retirement home?"

"After his success, Bellerophon grew very proud, and stupidly tried to call on the Olympians in their palace without having been invited. He rode through the air on Pegasus, dressed in crown and robes. Zeus asked Hera to send a gadfly to sting Pegasus under the tail! Pegasus reared, and Bellerophon tumbled off. Since then, he had been living under Zeus's curse."

"What happened to Pegasus?" I asked.

"Zeus caught Pegasus and used it as a pack-horse to carry thunderbolts."

Later in the day, I tried to work out the minimum number of spears Bellerophon must have prepared with lead in his fight against the Chimaera.

Solution

Observe that Bellerophon has to put either 11 units of lead down each of 7 jaws, or 7 units of lead down each of 11 jaws.

If the Chimaera shows up with 11 heads, there can be at most 7 units of lead on each spear, and he can have as many as 7 such spears. He cannot have 8. Otherwise, if the Chimaera shows up with only 7 heads, 2 such spears must go down the same jaw, and 3 units of lead will be wasted.

If the Chimaera shows up with 11 heads, each of these 7 spears needs to be supplemented by other spears with a total of 4 units of lead. Bellerophon can have as many as 4 spears each with 4 units of lead. He cannot have 5. Otherwise, if the Chimaera shows up with 11 heads, two of the 12 spears with at least 4 units on each must go down the same open jaw, and at least 1 unit of lead will be wasted.

The task can be accomplished with 17 spears. Add 3 spears each containing 3 units of lead, and 3 spears each containing 1 unit of lead.

The diagram below shows how these 17 spears are to be used. The upper row of boxes represent the situation when the Chimaera shows up with 7 heads, and the bottom row of boxes represent the situation when the Chimaera shows up with 11 head. The double solid lines represent spears with 7 units of lead, the single solid lines represent spears with 4 units of lead, the double dotted lines represent spears with 3 units of lead, and the single dotted lines represent spears with 1 unit of lead.

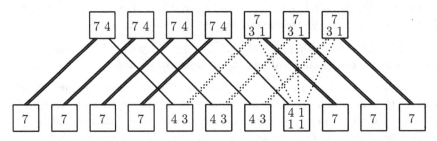

I must still show that 17 is the minimum number of spears. The diagram above has $7 + 11 = 18$ boxes. If the whole structure is linked together, such as the current case, we need at least 17 lines to connect them. In order to have a smaller number of connecting lines, the structure must fall into at least two separate parts.

In each part, the number of units of lead in each upper box must be 11, and the number of units of lead in each lower box must be 7. Hence the total number of units of lead in the part is a multiple of both 7 and 11. However, the least common multiple of 7 and 11 is 77, which means that the part contains all 77 units of lead. It follows that it is the only part.

Therefore, the whole structure must be linked together, and the minimum number of connected lines, or spears, is indeed 17.

Chapter 27

The Triangular Palace

"Another former hero, Eubie," said Silenus, "who had a Corinthian connection was Theseus. On a visit to Corinth, King Aegeus of Athens fell in love with Princess Aethra. She had got tired of waiting for Bellerophon, and they were secretly married. Theseus was their son. Aegeus left mother and son behind when he returned to Athens. His eldest nephew was expecting to be the next King of Athens, and would kill mother and son if he found out about their existence."

"When did Theseus come out from hiding, sir?"

"When he was 14, he went to his father. Aegeus had since married a witch named Medea. She was hoping that one of her sons would succeed to the throne. She knew by magic who Theseus was and tried to poison him. Luckily, Aegeus recognized his son in time, and knocked the cup which Medea was offering Theseus from her hand. The poison burned a large hole through the floor, and Medea vanished in a magic cloud."

"That was a narrow escape," I said.

"Then Aegeus's nephews ambushed Theseus on his way to meet the carriage which was sent to Corinth to fetch Aethra. Theseus fought and killed them all. The Athenians wanted Theseus punished."

"He was only acting in self-defense."

"The Athenians did not care. It happened some years before that Androgeus, the son of King Minos of Crete, was murdered by Aegeus's nephews. When Minos complained about this to the Olympians, they gave orders that Athens must send 7 boys and 7 girls every 9 years to be devoured by the Cretan Minotaur."

"Who or what was the Minotaur?" I asked.

117

"The Minotaur was a monster — half bull, half man — which Minos kept in a triangular palace surrounded by a labyrinth, built for him by a master Athenian craftsman named Daedalus."

"Surely, as the new heir, Theseus would not be put on this list," I said.

"Theseus himself asked to be put on it. He welcomed a chance to free his country from this horrid tribute. The ship in which he sailed carried black sails for mourning, but he took white sails along, too. He told his father that if he killed the Minotaur, he would hoist the white sails."

"He already had the air of a hero," I said in admiration.

"When he got to Crete, he prayed to Aphrodite. She sent your friend Eros to make Ariadne, Minos's daughter, fall in love with Theseus. That night, she offered to help him if he promised to marry her."

"Did he agree?"

"He did. Ariadne handed him a magic ball of thread, given her by Daedalus before he left Crete. Theseus tied the loose end of the thread to the labyrinth door, and the ball rolled by magic through all the twisting paths until it reached the triangular palace in the middle."

Silenus drew me a diagram of the palace, which was divided into 100 triangular rooms arranged in 10 rows in each of three directions.

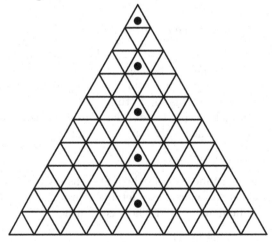

"Did the Minotaur sleep in one of the 5 rooms marked with black circles?" I asked.

"The Minotaur slept in any of the 100 rooms, a different one each night. To discover exactly which room, an observer must be in a room in the same row as the Minotaur's room, in any of the three directions. He could then creep up and kill the Minotaur while it was asleep. However, if 2 observers were in rooms in the same row, an alarm would be triggered. The Minotaur would wake up and kill them all."

"So if 5 observers occupied the rooms marked with black circles, that would work."

"Theseus wanted as many observers as possible, so that the distance between the Minotaur and the discovering observer could be reduced. The other 6 boys went in with Theseus. Ariadne offered to go in with them, but Theseus wisely declined."

"Where were the 7 observers stationed?" I asked.

"You can work it out later," said Silenus. "As it turned out, the Minotaur was sleeping in a room in the same row as Theseus. So he crept up and cut off the monster's head."

"Hurrah!" I shouted.

"The 7 boys followed the thread back to the entrance of the labyrinth. Meanwhile, Ariadne had freed the 7 girls, too, and together they sailed for Athens with the Minotaur's head."

"Was Theseus happily married to Ariadne?"

"That was not meant to be. When they stopped on the island of Naxos, Dionysus suddenly appeared and wanted to marry Ariadne himself."

"Did Theseus fight Dionysus for her?"

"Theseus dared not offend Dionysus, and he had no great love for Ariadne anyway. Ariadne decided that it would be far more glorious to marry a god than a mortal. So they parted amicably. There was one other twist of fate. In the excitement, Theseus had quite forgotten to change the sails. Aegeus, watching anxiously from a cliff at Athens, saw the black sails appear instead of the white. Overcome by grief, he jumped into the sea and drowned. This tragedy made Theseus King of Athens. He then made peace with the Cretans."

It was a wonderful story, full of twists. Then I settled down and worked on the problem of posting observers inside the Minotaur's palace.

Solution

It took me quite a while, but I finally found a way of posting Theseus and the other 6 boys without 2 of them in rooms in the same row. I showed Silenus my diagram.

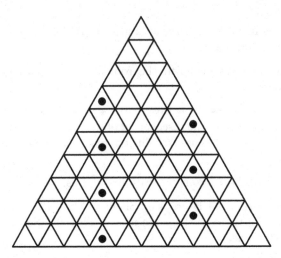

He was pleased, but wanted more. He asked me whether it was indeed wise for Theseus to refuse adding Ariadne as an extra observer.

I set up a coordinate system for the rooms as follows. I labeled the sides of the large triangle X, Y and Z as shown in the diagram below. Each room was assigned coordinates (x, y, z) such that $x = k$ if and only if the room was in the kth row from side X, and so on. The diagram illustrated the case where there were three rows in each direction, and nine rooms altogether.

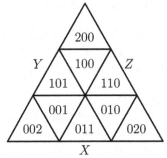

In the actual palace, the corner rooms were assigned coordinates $(9,0,0)$, $(0,9,0)$ and $(0,0,9)$. For any room in the shape of a right-side-up triangle, the sum of the three numbers in its label was 9. For a room in the shape of an upside-down triangle, the sum of the three numbers in its label was 8.

"Suppose, sir," I explained, "that 8 observers were placed in the rooms labeled (x_1, y_1, z_1), (x_2, y_2, z_2), ..., (x_8, y_8, z_8). If no two xs were the same, no two ys were the same and no two zs were the same, then the sum of all the xs, ys and zs was at least $3 \times (0 + 1 + 2 + 3 + 4 + 5 + 6 + 7) = 84$. On the other hand, the sum $x + y + z$ for each room was at most 9, yielding a total of at most $9 \times 8 = 72$. This contradiction meant that two of the xs, two of the ys or two of the zs must be the same, and the Minotaur would wake up."

"So you see, Eubie," said Silenus, "had Theseus accepted Ariadne's offer, the story would have a totally different ending."

"By the way," I asked, "how did Theseus fall from grace?"

"He became conceited just like Bellerophon. Instead of visiting Olympus uninvited, he went down to Tartarus and challenged Hades to a wrestling match. Hades told him to sit down on a bench and wait for Hades to change clothes. The bench was a magic one, and Theseus became attached to it."

"How did he manage to escape and come to Perseus in Tiryns?"

"He stayed there for a very long time, until he was yanked off the bench by your friend Heracles while performing his twelfth labor. Theseus had no more adventures after that."

Chapter 28

The Matching of Boxes
and Keys

"Please tell me, sir," I said to Silenus, "something about this Daedalus who built the labyrinth for Minos's Minotaur."

"He is an Athenian, Eubie," replied Silenus, "as I had mentioned. He is a wonderfully skillful smith. When he was a boy, he was taught by Athene and Hephaestus. He once worked for Minos, who was a slave driver and would not give him any holidays. So Daedalus decided to escape."

"Crete being an island, did he steal a boat and sail away?"

"No, Minos's fast ships would soon have overtaken him. So he made himself and his son Icarus a pair of wings each, to be strapped on their arms. The big quills were threaded to a frame, and the smaller feathers were held together by bees-wax."

"That was clever of him."

"Unfortunately, it ended in a disaster. Having helped Icarus on with his pair of wings, Daedalus warned his son not to fly too low, for fear of the sea spray, or too high, for fear of the sun."

"Since you said it was a disaster, did Icarus disobey his father?"

"Most people would come to that conclusion, but I happen to know that it was not true. Do you remember Orion's friend Phaëthon, the son of Helius the Sun? Every morning, Helius harnesses four white horses to a fiery chariot across the sky from Colchis to another palace of his in the Far West. There he unharnesses the horses, and when they have grazed, loads them and the chariot on a boat which sails round the world by way of the Western Ocean, reaching Colchis again the next morning."

"When does Helius sleep?"

"Through the night while the boat is sailing. He can never take a holiday from work."

"How does Phaëthon come into the story?"

"The boy was always pressing his father for permission to drive the chariot. On that fateful day when Daedalus and Icarus took off on their wings, Helius finally relented. However, Phaëthon was unable to manage the reins, and the chariot started plunging up and down. The Olympians felt icy cold one minute, and scorching hot in the next. Zeus ordered Phaëthon to put things right at once, but the boy could not do so. In disgust, Zeus threw a thunderbolt and killed him."

"Poor Phaëthon, and poor Helius. What happened to Icarus then?"

"When the chariot was out of control, Daedalus adjusted his altitude quickly, but Icarus was not skillful enough to follow suit. When it got too hot for him, the wax melted and the feathers came loose. He lost height, fell into the sea and drowned. He was buried on a small island, later called Icaria, where the sea had washed up his body."

"That was sad. What happened to Daedalus then?"

"He flew to Sicily and sought refuge in the court of King Cocalus. He gave the king's daughters a set of beautiful dolls, with moveable arms and legs. They were so delighted that they persuaded their father to let Daedalus stay. When Minos pursued Daedalus by ship, Cocalus denied having seen him."

"Good of him," remarked Heracles.

"However, the crafty Minos offered a big reward for a challenge. The task was to pass a thread through all the holes of a seashell. Cocalus was greedy and wanted the reward, and asked Daedalus for help. To repay Cocalus's kindness, Daedalus solved the problem for him."

"How?"

"He tied the thread to a hind-leg of an ant and put the ant into the shell. Then he smeared honey around the holes one at a time. To eat the honey, the ant went through the seashell, pulling the thread behind it."

"That was really clever."

"Minos thought so too. After paying the reward to Cocalus, he declared that only Daedalus could have thought of that! He threatened to burn down Cocalus's palace unless he gave up Daedalus. His daughters persuaded him not to yield to Minos's demand. However, to save the palace, Daedalus advised

Cocalus to concede that Daedalus had indeed been to Sicily, and had left behind a document torn into 100 pieces. Each was put inside a locked box with a unique key, and if all pieces were retrieved, they could be put together to reconstruct the document. It contained information about where Daedalus would be hiding."

"What was the point if Cocalus handed over all the boxes as well as all the keys?"

"The boxes and the keys were numbered 1 to 100 respectively, but the number of the correct key might differ from the number of the box by at most 1. For instance, you may start with trying key 1 on box 1. If it opens, try key 2 on box 2. If all goes well, you know that key 100 must open box 100. So you can find out which key goes with which box in 99 attempts."

"Suppose it goes well until key 42 does not open box 42," I said, catching on the idea, "then I know that box 42 must be opened by key 43 while key 42 must open box 43. This actually saves me an attempt."

"Excellent," said Silenus. "So with good luck, you can do it in 50 attempts."

"How many attempts did Minos make?"

"None," said Silenus. "Before he started, Cocalis's daughter invited him to take a warm bath in the new bathroom built by Daedalus. They poured boiling water down the bathroom pipe instead of warm water, and scalded Minos to death. Cocalus pretended that this was an accident, blaming a faulty pipe for blocking off the cold water which was to be added."

"However stupid they might be," I said, "the Cretans could not possibly buy that."

"Of course not," replied Silenus. "However, Minos had not taken along anyone in a position to make major decisions. It was even possible that whoever succeeded to power back in Crete might be grateful to Cocalus for getting Minos out of the way. In any case, a distant war would not be in the interest of Crete. So the Cretans pretended to believe the story and dropped the matter."

"It was really hard luck on Minos."

"Upon his arrival at Tartarus, he was appointed one of three Judges of the Dead. Zeus, his father, had made an arrangement with Hades."

I pondered over this sad story for a while. Then I returned to the problem of the boxes and the keys. I was convinced that 99 attempts were unnecessary, even when everything was against me. I wondered what the minimum number of attempts I would need to guarantee the complete matching of boxes and keys.

Solution

I first reduced the number of boxes and keys from 100 down to 4, and came up with the following plan. In the first attempt, I would try key 3 on box 3. There were two cases.

Case 1. The attempt was successful.

Then I knew that key 4 would open box 4. In the second attempt, I would try key 1 on box 1. If successful, I would know that key 2 would open box 2. Otherwise, key 1 would open box 2 and key 2 would open box 3.

Case 2. The attempt failed.

In the second attempt, I would try key 3 on box 2. If successful, then key 1 must open box 1 and key 2 must open box 3. Suppose the attempt failed. Then key 3 must open box 4, while key 4 must open box 3. Now a third attempt was needed to decide whether key 1 would open box 1 or box 2.

So 3 attempts were sufficient. A nagging doubt was whether there was a way to get it done in two attempts.

Instead of trying to discover further tactics, I decided to look at the strategic picture. There were 5 possible scenarios, as shown in the following chart.

Scenario	Box 1	Box 2	Box 3	Box 4
First	Key 1	Key 2	Key 3	Key 4
Second	Key 2	Key 1	Key 3	Key 4
Third	Key 1	Key 3	Key 2	Key 4
Fourth	Key 1	Key 2	Key 4	Key 3
Fifth	Key 2	Key 1	Key 4	Key 3

Now each attempt could result in 2 different outcomes: success or failure. Thus 2 attempts could distinguish among at most 4 possible scenarios. Hence 3 attempts were indeed necessary.

Returning to the actual problem, I divided the 100 boxes and keys into 25 sets of 4. There was now the possibility that box-key interchanges might occur between sets. However, in the algorithm above, there were times when less than three attempts were required. I went over it again for the first set, with the additional complication that box 5 and key 5 might be involved.

In the first attempt, I would try key 3 on box 3. There were two cases.

Case 1. The attempt was successful.

In the second attempt, I would try key 1 on box 1. If successful, I would know that key 2 would open box 2. Otherwise, key 1 would open box 2 and key 2 would open box 3. I could now try key 4 on box 4 in the third attempt.

Case 2. The attempt failed.

In the second attempt, I would try key 3 on box 2. If successful, then key 1 must open box 1 and key 2 must open box 3. I could now try key 4 on box 4 in the third attempt. Suppose the second attempt failed. Then key 3 must open box 4 while key 4 must open box 3. Now a third attempt was needed to decide whether key 1 would open box 1 or box 2.

The maximum number of attempts for each set of 4 remained 3, so that $3 \times 25 = 75$ attempts were sufficient. The minimum number of attempts for each set of 4 also remained 3. Thus 75 attempts were necessary.

Chapter 29

The Number-and-Coin Tricks

Triptie's mission took him home for the first time. My mother was delighted to see how tall he had grown. He was taller than when he and I were touring the Sporades Archipelago together.

My father was planning to hold a state dinner in Triptie's honor. Triptie wanted to give some sort of performance as an after-dinner entertainment. However, his voice was hoarse and his movement was awkward. A song-and-dance was clearly out of the question.

He decided to perform some tricks. For days, he was closeted with Silenus. Finally, they were ready to test it on me.

"I will be out of the room at the beginning of the performance," Triptie said. "You and the audience will choose a number and place several coins in a row on the table, each either heads up or tails up. Silenus will remain in the room, and will know what number has been chosen. He will either do nothing or flip over exactly one coin. Then I will come in, look at the row of coins, and announce that number."

"From what range can the number be chosen, and how many coins will be placed?"

"Let's start with a simple example, Eubie," Silenus cut in. "In Trick Number One, the range is from 0 to 1, and the number of coins is 1."

"That is not going to impress anyone, sir," I said. "You can decide in advance that if the coin is heads up, then the number is 0, and if it is tails up, then the number is 1. If the coin is telling the truth, you do nothing.

Otherwise, you flip it. In either case, Triptie can tell what number has been chosen."

"Perhaps that is too simple," conceded Silenus. "In Trick Number Two, the range is from 0 to 3, and the number of coins is 3."

"This is still no big deal," I said. "If all three coins are the same, the number is 0. If the first coin is different from the other two, then the number is 1. If the second coin is different from the other two, then the number is 2. If the third coin is different from the other two, then the number is 3. No matter how the coins are placed initially, you can make them tell the truth by flipping at most one coin."

"You are pouring cold water over my plans, little brother," complained Triptie.

"The average audience will not be as smart as your little brother," Silenus comforted Triptie. "We can still perform Trick Number Two. If needed, we will have a grand finale. In Trick Number Three, the range is from 0 to 7 and the number of coins is 7."

"Isn't it still the same thing?" I questioned. "If all the coins are the same, the number is 0. Otherwise, whichever is different tells you what non-zero number it is."

"What would I do, as the assistant, if the audience places three coins heads up and four coins tails up?" asked Silenus. "Remember, I can flip at most one coin."

"Oops!" I gulped. "I hadn't thought of that."

I thought hard about it on the days before the gala event, but could not see through the trick. After the state dinner, my father stood up and said. "Citizens of Eleusis, I now present Crown Prince Triptolemus who will perform a trick with the assistance of Silenus, the Royal Tutor."

During the round of applause, Triptie took a bow and left the dining room. Silenus explained Trick Number Two to the audience. The audience chose the number 2, and placed three coins heads up, heads up and tails up in that order. Silenus flipped over the first coin and summoned Triptie, who came in and announced the number to the amazement of the audience.

This was repeated a few times, with guaranteed success. Silenus and Triptie decided to leave well enough alone and called it a day.

I continued to try to figure out Trick Number Three after Triptie had left to resume his mission. Unable to make any progress, I sought help from Silenus.

Solution

"Eubie," he said, "your approach to Trick Number Two is brilliant. It is different from the solution Triptie and I had in mind. Actually, our idea is harder to understand than yours. Our advantage is that the argument can be generalized to work for Trick Number Three, whereas you have seen that your approach fails in that regard."

"What is your solution, then?"

"We consider all combinations of two symbols, which we call A and B. There are four in all: neither A nor B, A only, B only and both A and B. We denote these by (), (A), (B) and (AB), respectively. We agree that () means 0, (A) means 1, (B) means (2) and (AB) means 3."

"It is more or less like saying A is 1 and B is 2. The sum is 0 for (), 1 for (A), 2 for (B) and 3 for (AB)."

"Good observation," Silenus nodded with approval.

"However, there are only three coins. How do they correspond to the combinations?"

"Let us *associate* the first coin with A, the second coin with B and the third coin with AB. The following chart shows all 8 possible placements of the coins, with 0 standing for heads up and 1 standing for tails up."

A		A		A		A		A		A		A		A
	B	B			B	B			B	B			B	B			B	B			B	B
0	0	0		0	0	1		0	1	0		0	1	1

A		A		A		A		A		A		A		A
	B	B			B	B			B	B			B	B			B	B			B	B
1	0	0		1	0	1		1	1	0		1	1	1

"How does this help?"

"The key concept is *parity*, that is, the state of being odd or even. In each placement, there are two coins associated with A, namely the first and the third, and two coins associated with B, namely the second and the third. We will include A if the sum of the two numbers in the columns with A is odd, but exclude A if the sum of the two numbers in the columns with A is even. The inclusion or exclusion of B follows the same rule."

"Let me see. In the 8 cases, the results are (), (AB), (B), (A), (A), (B), (AB) and (), respectively. Am I right?"

"You are. Suppose the audience chooses the number 2. What would I do, as the assistant, in each case?"

"In the third and the sixth cases, you do nothing because the coins are already telling the truth. In the second and the seventh cases, you flip the first coin. In the first and the eighth cases, you flip the second coin. In the fourth and the fifth cases, you flip the third coin."

"You have caught on to the idea. Since there are two cases for each of (), (A), (B) and (AB), there is always a way to reach one of them by flipping at most one coin."

"For Trick Number Three, I assume that you add a third symbol C, so that there are 8 combinations, namely, (), (A), (B), (AB), (C), (AC), (BC) and (ABC), meaning 0, 1, 2, 3, 4, 5, 6 and 7, respectively."

"Yes," Silenus agreed. "Let's take an example and see what happens. Suppose the audience chooses the number 5 and places the coins as shown in the following chart."

A		A		A		A
	B	B			B	B
			C	C	C	C
0	1	0	0	0	1	1

"Let me see. The number 5 corresponds to (AC). This means that the sum of the 4 numbers in the columns with A is odd, those with B is even and those with C is odd. Currently, the coins are telling the truth as far as A is concerned, but not so for either B or C. Obviously, the way to change the parity for both B and C by flipping only one coin is to flip the coin associated with (BC), yielding the following chart."

A		A		A		A
	B	B			B	B
			C	C	C	C
0	1	0	0	0	0	1

"You are right. When Triptie comes in and looks at the coins, he can check that the sums for A, B and C are odd, even and odd respectively. Hence the combination is (AC), which corresponds to the number 5."

Chapter 30

The Unguarded Gate

The 5 months Persephone enjoyed above the ground were coming to an end. Demeter took her to Eleusis and handed her over to Hades. There was a tearful farewell, and my mother did not bake any fruit cakes this time.

Demeter told me that I had been invited to a wedding at the Mount Olympus Inn. I was surprised because I knew of no one there who was to be married. Demeter also said that I should stay for a while afterwards, because everybody missed me, especially Hebe, Heracles and Eros. My parents gave their eager consent, and awarded Silenus another vacation. By now, I was an experienced traveler, and there was no need for Silenus to accompany me.

Demeter's chariot landed in the foothills and we went to the Mount Olympus Inn. The happy couple turned out to be Admetus and Alcestis of Pherae. They were renewing their marriage vow.

Apollo was there too, as we expected. He gave Admetus a wedding present. He arranged with Hades that when Admetus's last day arrived, someone else might go down to Tartarus instead of him. Hades did not care which ghost he got, as long as it came willingly and punctually.

After the wedding banquet, we went up to the palace. My full pass from last time was still good. Demeter dropped me off at the gatehouse. Only Hebe was home. Heracles had been called away because Argos and Thebes were at war, and Thebes was under siege.

The next day, Hebe invited Eros over to keep me company. He was delighted. The only other person of his age in the Palace was Zeus's cup-bearer Ganymede, who also lived in the Creamy Way. However, they did not get along. Typical of the residents there, they quarreled whenever they played together.

Heracles came back two days later. We asked him to tell us the whole story.

"When the throne of Polyneices, King of Thebes, was usurped by his brother Eteocles, Polyneices sought refuge with his friend Adrastus, King of Argos. Amphiaraus, the husband of Adrastus's sister Eriphyle, advised against it."

"Why would that lead to war, sir?" I asked. "Wouldn't Eteocles be pleased?"

"Adrastus did not follow the advice of Amphiaraus, Eubie. Polyneices bribed Eriphyle to persuade Adrastus not only to allow Polyneices to stay, but help him regain his throne by sending an army to attack Thebes."

"I assume," I said, "that you went to fight for Thebes. Since you came back, you must have succeeded in beating back the Argos army."

"What I am telling you, Eubie, is about the First Siege of Thebes which happened many years ago. What just happened was the Second Siege of Thebes."

"What happened last time?" asked Hebe.

"The Argos army was formed into 7 companies, to attack all 7 gates of Thebes at the same time. The siege was long and bloody. Many died on both sides, including 4 of the 7 Argos company leaders. Finally, Polyneices offered to fight Eteocles in a duel for the throne. Eteocles accepted the challenge, and they ended up killing each other. Adrastus decided to retreat, but now the Thebans gave chase. Amphiaraus died when his chariot fell into a ravine. So Adrastus was the only company leader who survived."

"What brought on the Second Siege of Thebes?" Eros asked.

"The sons of 5 of the 6 dead company leaders wanted revenge, except for Amphiaraus's son Alcmaeon, whose was just as sensible as his father. Polyneices's son bribed Eriphyle. She persuaded Adrastus to attack Thebes again. Adrastus, already an old man, sent in his place his son and heir to the throne of Argos."

"Since you have come back, my dear husband," said Hebe, "the Thebans must have beaten off the Argives once again."

"This is true up to a point. The Argives were indeed beaten off, and were prepared to retreat the next day. This time, only 1 of the 7 leaders died, Adrastus's son. Paradoxically, this led to the fall of Thebes."

"How?" I exclaimed.

"It had been prophesied that Thebes would never fall as long as Adrastus lived. The Thebans knew that he was bound to die from grief when he heard of the death of his son. However, they believed erroneously that the fall of Thebes would necessarily follow from the death of Adrastus. As the siege was being lifted at night, many Thebans stole away. These included fighting men, leaving behind a skeleton force to guard the 7 gates."

"Did Thebes fall then?" asked Hebe.

"Not just then," said Heracles. "That was when I arrived on the scene. I organized the remaining fighting men into several squads that night, with at least 2 squads guarding each gate initially."

"Did they get to sleep at all?" Eros asked.

"They only needed to be on guard until the morning, when the Argives were expected to leave. To keep them from falling asleep, I divided the night into 10 segments of equal length, by having a watchman count pebbles at a constant rate. I gave the order that after every interval, the watchman would set off a drum-roll. Each squad would then move to the next gate."

"Did they move to the left or to the right?" I asked.

"This was where I faltered. I should have designated at least one squad from each gate to move in the same direction. Then no gate would be left unguarded. However, things were chaotic, and I did not have time to go around the wall to give orders to specific squads. I let each squad choose independently the direction in which it would move. However, once they had made the choice, they would stick to it. In other words, a squad moving to the left always moved to the left, and a squad moving to the right always moved to the right."

"How did Thebes fall, then?" Eros asked.

"Some time after things had settled down, I went around the wall and checked on things. I found exactly 4 gates with exactly 1 squad over each of them. I did not appreciate then the full implication of that piece of information. Later, the Argives spotted an unguarded gate, broke into Thebes and plundered the city."

"How did you managed to get away, my dear husband?"

"All was lost. The best I could do was to organize a sortie through the gate that was attacked by the son of Adrastus, saving as many people trapped in the city as I could."

Heracles was getting tired. Hebe went and prepared a warm bath for him. She offered Eros and me some barley cakes. Eros decided to go home for lunch instead. As I ate, I began to ponder why that observation of Heracles should have warned him that a gate was unguarded.

Solution

I called the squads moving to the left black squads and those moving to the right white squads. Initially, there were at least 2 squads over each gate. Suppose at that time, there was a squad of the same color, say black, over every gate. Then there would always be a black squad over every gate, because these 7 just rotated around the wall.

Initially, there was another squad over each gate. Among these 7, 4 had to be of the same color. They would always be over different gates, meaning that at any time, there would be four gates with at least 2 squads, and at most 3 gates each with at most 1 squad.

However, this was contrary to Heracles's observation. I concluded that initially, there was a gate A with no black squads, as well as a gate B with no white squads. That was as far as I got when Hebe joined me for lunch. While munching on the barley cakes, she came up with a brilliant idea.

"All these movements are confusing," she said. "Let us pretend that the black squads are stationary while the white squads move to the right but skip over exactly one gate after each interval. Then gate A will never have black squads. There will always be a gate without white squads, shifting from gate B to the left 2 gates per interval. Since 7 is an odd number, it will shift onto gate A within 7 intervals. At that point, gate A will have no squads at all."

"That solved the problem," I exclaimed, "because the squad distribution in your hypothetical scenario is exactly the same as the actual one. Well done!"

Chapter 31

The Rearrangement of the Suitresses

Heracles missed the wedding of Admetus and Alcestis, but now had a funeral on his hands. He went down to the Mount Olympus Cemetery in the company of Zeus, who was to give the eulogy. Hebe and I went along.

"When Narcissus was born," said Zeus, " it was prophesied that he would live to a ripe old age if he never saw his own image. His mother kept no mirrors in the house, and washed his face for him all the time. When he grew up, he kept hearing others saying how handsome he was. He began to fall in love with himself. Many girls wanted to marry him, but he was not interested in any of them, and sent them all away. Finally, he saw his own reflection in a pool, bent down for a kiss, fell in, and drowned."

He did not seem a very interesting person, and his life seemed unremarkable. I wondered why Zeus would deliver the eulogy himself. So I asked Heracles about it.

"My father did not really know Narcissus, Eubie," said Heracles. "He gave the eulogy because of Echo, a nymph who loved Narcissus. Although my father was not interested in Echo herself, he used her to make liaisons with other nymphs, behind Hera's back. One time, Hera pressed Echo for information when she missed Zeus. Disguised as a swan, Zeus was having an affair, not with another nymph but with Queen Leda of Sparta. Echo knew about it, and covered for Zeus by telling Hera that she had seen Zeus sneaking away, disguised as a woodpecker. So Hera started catching woodpeckers, and found each of them just an ordinary girl. Then she realized that she had been fooled."

"Did Hera kill Echo, sir?" Hebe asked.

"No, Echo was made invisible. Moreover, she could not say anything of her own but could only repeat the final words of other people's speeches. She fell in love with Narcissus, but her current state prevented her from declaring her love. Narcissus went out hunting with other boys one day, and got separated from his companions. Echo followed him. He heard footsteps close by, but saw nobody."

I could imagine the ensuing conversation.

"Is anyone here?" Narcissus shouted.

"Here," answered Echo.

"Then come to me!" He mistook her for one of his friends.

"Come to me!" she answered.

"Here I am!"

"I am!"

"Did she catch up with Narcissus?" asked Hebe.

"She did. She rushed to Narcissus and threw her arms around his neck. Narcissus was horrified that he was being kissed by a girl. He broke away and ran home. Echo made several other attempts to approach Narcissus, but all to no avail. Though we cannot see her, I am convinced that she is here, attending the funeral."

After hearing all this, I still felt that Narcissus was not a very interesting person, and his life was unremarkable. Heracles agreed with me on the whole. Then he told me an episode which piqued my interest.

Many girls wanted to marry Narcissus. His mother chose 20 of them and had them line up in a row before Narcissus. They were numbered from 1 to 20 from left to right. Narcissus applied delaying tactics. He requested that the girls be rearranged in the opposite order. In other words, they should be numbered from one to twenty from right to left. Two girls standing next to each other might exchange positions, but this had to be supervised by his mother, meaning that such exchanges could take place only one at a time.

His mother complained that this would take far too long. Narcissus made a concession, allowing two girls whose positions, not their numbers, differ by a multiple of 4 to exchange places without supervision. In other words, the girls in positions 1, 5, 9, 13 and 17 could change places any time. The same applied to the girls in positions 2, 6, 10, 14 and 18, those in positions 3, 7, 11, 15 and 19, as well as those in positions 4, 8, 12, 16 and 20.

The girls made a lot of unnecessary exchanges. By the time the rearrangement was completed, it was time for lunch. Afterwards, the girls lined up once again where they left off before lunch, numbered from 1 to 20 from right to left. Now Narcissus wanted them back in the original order, numbered from 1 to 20 from left to right.

The rule for the supervised exchanges remained the same. Unsupervised exchanges were now allowed between two girls whose positions, not their numbers, differed by a multiple of 5. In other words, the girls in positions 1, 6, 11 and 16 could make exchanges any time. The same applied to the girls in positions 2, 7, 12 and 17, those in positions 3, 8, 13 and 18, those in positions 4, 9, 14 and 19, as well as those in positions 5, 10, 15 and 20.

More unnecessary exchanges were made. By the end of the day, the rearrangement had not been completed. Tired and frustrated, the girls decided to go away and never come back.

After the funeral, Heracles stayed behind to tidy things up while Hebe and I returned to the gatehouse. The question that came immediately to our minds was what the minimum number of supervised exchanges was in the morning, and what it was in the afternoon.

Solution

"Let us start with the morning problem," I began, "and cut the number of girls down to 4."

Hebe objected, "That is going too far, because no unsupervised exchanges are now possible. Let us cut it down to 8. To make the rules clearer, we mark the 8 positions with colored circles, red, yellow, blue, green, red, yellow, blue and green. Then two girls on circles of the same color could exchange positions unsupervised, while two girls on circles of adjacent colors could only exchange under supervision."

"By your scheme," I said, "all girls on red circles must go to green circles, and vice versa. All girls on yellow circles must go to blue circles, and vice versa. Since yellow and blue are adjacent colors, we can use two supervised exchanges to rearrange the girls on them. Similarly, the four girls on red and green circles can be sorted out using another two supervised exchanges, for a total of 4."

I prepared the following chart, and got stuck.

Exchange Type	Girl Numbers							
(At the Start)	1	2	3	4	5	6	7	8
Supervised	1	3	2	4	5	6	7	8
Supervised	1	3	2	5	4	6	7	8
Supervised	1	3	2	5	4	7	6	8
Unsupervised	1	7	2	5	4	3	6	8
Unsupervised	1	7	6	5	4	3	2	8

Hebe came to the rescue and continued my chart.

Exchange Type	Girl Numbers							
Unsupervised	4	7	6	5	1	3	2	8
Unsupervised	4	7	6	8	1	3	2	5
Supervised	4	7	6	1	8	3	2	5
Unsupervised	8	7	6	1	4	3	2	5
Unsupervised	8	7	6	5	4	3	2	1

"With 20 girls," she continued, "we can use 9 supervised exchanges to get them into the order 1, 19, 18, 17, ..., 3, 2 and 20. Now exchange 1 with 16 and 15 with 20 without supervision. Then do a supervised exchange between 1 and 20, followed by unsupervised exchanges between 16 and 20, as well as between 1 and 15. Hence 10 supervised exchanges are sufficient."

It was now time for lunch.

"Are 10 supervised exchanges necessary?' I asked after we refreshed ourselves with a drink of nectar.

"Yes," said Hebe. "As you have already pointed out, all 5 girls on red circles must go to green circles, and all five girls on green circles must go to red circles. Unsupervised exchanges do not help here. So we need at least five supervised exchanges. We need another 5 for the 10 girls on yellow and blue circles. Hence $5 + 5 = 10$ is indeed the minimum number. Let us move onto the afternoon problem."

Based on our earlier success, we decided to mark the positions by circles in five colors, red, yellow, purple, blue and green, repeated in the same order. Then the girls on purple circles must remain on purple circles. Girls on red circles must move to green circles, and vice versa. The same applied to girls on yellow and blue circles. The new complication was that yellow and blue were no longer adjacent colors.

Since red and green were still adjacent colors, we could use 4 supervised exchanges to put the girls these circles to circles of the correct colors. The same task for yellow and blue would require 8 supervised exchanges because each girl on such a circle must go through two exchanges to get to a circle of the other color. However, these girls could not get there directly without taking part in at least 1 exchange involving a girl originally on a purple circle. So the minimum was at least 13.

"Would 13 be the actual minimum?" I wondered.

"I think so," said Hebe. "After using 4 supervised exchanges to sort out the girls on red and green circles, we can proceed as follows."

Exchange Type	Girl Numbers			
(At the Break)	2 3 4	7 8 9	12 13 14	17 18 19
Supervised	**3 2** 4	7 8 9	12 13 14	17 18 19
Supervised	3 **4 2**	7 8 9	12 13 14	17 18 19
Unsupervised	3 8 2	7 **4** 9	12 13 14	17 18 19
Supervised	3 8 2	**4 7** 9	12 13 14	17 18 19
Supervised	3 8 2	4 **9 7**	12 13 14	17 18 19
Unsupervised	3 8 2	4 **13** 7	12 **9** 14	17 18 19
Supervised	3 8 2	4 13 7	**9 12** 14	17 18 19
Supervised	3 8 2	4 13 7	9 **14 12**	17 18 19
Unsupervised	3 8 2	4 13 7	9 **18** 12	17 **14** 19
Supervised	3 8 2	4 13 7	9 18 12	**14 17** 19
Supervised	3 8 2	4 13 7	9 18 12	14 **19 17**
Unsupervised	**14** 8 2	4 13 7	9 18 12	**3** 19 17
Supervised	14 8 2	4 13 7	9 18 12	**19 3** 17
Unsupervised	**19 18 17**	**14 13 12**	**9 8 7**	**4 3 2**

Chapter 32

The Wily Prisoners

Apollo returned to the Mount Olympus Palace with no warning. He came to the gatehouse and found Hebe and me working on problems.

"Where is Heracles?" he shouted.

"I think he is down in the foothills on some errand for father," Hebe replied. "What do you want him for?"

"I don't exactly want him. I want Dionysus, but I can't find him. If anyone knows where he may be, Heracles would know."

"I had seen Dionysus around earlier in the day," I said. "I will help you look for him."

In the end, Apollo himself found Dionysus. He had gone to the Creamy Way to visit his mother. Apollo got some wine from him.

"What do you want the wine for?" Dionysus asked.

"I want the Fates to try some. I am sure they will like it. You see, they are about to snip off the life-line of my friend and master Admetus, King of Pherae. Uncle Hades has promised me that Admetus can find a substitute to take his place. However, that might take some time. So he sent me here to see if I could delay the Fates."

"Good luck," said Dionysus. "I will be very happy if the Fates like wine. That will make it acceptable to even more people."

The Fates got drunk, buying Admetus valuable time. Apollo at once returned to Pherae to bring Admetus the good news, and to see whether he had found anyone willing to take his place.

Later in the day, Apollo was back again. This time, he found Heracles in the gatehouse. He told the three of us the story.

First, Admetus went to his parents, who were nearly 100 years old. "Will either of you die for me?" he asked them.

"Certainly not! What kind of a son are you, asking your parents to die for you? We may be old, but we can still enjoy life."

Then he went to a poor man suffering from an incurable disease, and asked him, "Will you die for me?"

"Certainly not! You may think that my disease is incurable, but there is always hope. Another Asclepius might come along and save my life."

King Admetus summoned 11 prisoners in the dungeon. He said to them, "If one of you will die for me, I will set the others free. Otherwise, I will have all of you executed at once. Will any of you die for me?"

"Certainly not!" they said together. "Pherae is a democracy and you cannot execute us without due process. We hope you will die soon. The new king may proclaim an amnesty and release us all."

Admetus devised a plan. Each prisoner would have 10 dots painted in a row across his forehead. Each dot was either black or white. Every prisoner could see the dots on the foreheads of the others, but not those on his own. A prisoner would die if he could not name correctly the color of each dot on his forehead.

After mutual consultation, the prisoners asked for one condition. They would be seated in a circle while the dots were painted. After they had the chance to examine those on the foreheads of the others, each might put up either his left hand or his right hand, and all hands would be put up simultaneously. Then each would name the color of each dot on his forehead.

Admetus readily agreed to this, not seeing how this would help the prisoners. However, he was surprised when they all gave the correct answers.

"At that point, Hermes came for Admetus. The Fates had become sober again, and unless Hermes could take someone else immediately, Admetus must go with him."

"So in the end," said Heracles, "he died."

"He did not," said Apollo. "As a last resort, he turned to Alcestis, with whom he had just renewed the marriage vow. She had not produced any children until the last couple of years, and now has two young sons. She agreed to die for him, kissed her sons farewell and took some poison."

"What a despicable person this man is," exclaimed Hebe.

"Somehow," I said, "I sensed that this is not the end of the story either."

"You are right, Eubie," Apollo confirmed. "When Alcestis got to Tartarus, she was met by Persephone. Persephone refused to admit Alcestis. She did not think that women should sacrifice themselves for their selfish husbands. She had a heated argument with Uncle Hades. In the end, Admetus sacrificed a pig to Uncle Hades, using the pig's soul to take the place of his own."

"Would Hades agree to that?" I asked.

"Uncle Hades disliked the exchange, naturally, but he did not want to continue the argument with Persephone. So he accepted the pig's soul, claiming that it was worth more than that of Admetus!"

"And he was right," remarked Heracles.

"Whatever made Alcestis choose to die for Admetus?" asked Hebe.

"I wondered about that too," said Apollo. "She explained that if Admetus had died, his brother would have seized the throne and murdered her young sons."

After Apollo had left, Hebe and I began working on how the wily prisoners had outwitted Admetus.

Solution

"Let us agree that the number of dots is 1 less than the number of prisoners," I said. "So with 2 prisoners, each has only 1 dot on his forehead. A simple system is for you to raise the right hand if you see a white dot, and raise your left hand if you see a black dot. So there is no problem here."

"With 3 prisoners," said Hebe, "each has 2 dots on his forehead, so that there are 2 decisions to be made. On the other hand, each receives 2 pieces of information, the hand-raising of each of the other 2 prisoners."

"Let us call the prisoners A, B and C," I said. "A will declare that his first dot is white if B raises his right hand, and black if B raises his left hand. A will declare that his second dot is white if C raises his right hand, and black if C raises his left hand."

"I don't think that would work," Hebe said. "You have only shown that it works for A. What about B? By your system, B will declare that his first dot is white if C raises his right hand, and black if C raises his left hand. However, whether C raises his right hand or his left hand is already determined by the color of the second dot on A's forehead. Unless the second dot of A and the first dot of B have the same color, the system will fail for B."

"You are right," I said. "How C decides which hand to raise must depend on the colors of the second dot of A as well as the first dot of B. I have an idea. Raise the right hand if the two have the same color, and the left hand if they have different colors. Since A can see the color of B's first dot and B can see the color of A's second dot, things will work out."

"Good, but not good enough," said Hebe. "With 4 prisoners, each will have 3 dots on his forehead. Each hand-raising decision is based on the colors of 1 dot on each of the other 3 prisoners. You can no longer talk about the same or different, because if not all 3 are of the same color, there will be six possibilities to be considered."

We were stuck, but not for very long. We both came up with the same idea. Going back to the case with 3 prisoners and 2 dots, to say that the 2 dots had the same color meant the that number of white dots among them was even, either 0 or 2. To say that they had different colors meant that the number of white dots among them was odd, namely, 1.

The parity concept could be extended to any number of prisoners and the corresponding number of dots. I constructed an example and showed it to Hebe.

Prisoner	Colors of Dots										Hand
#1	W	B	B	W	B	B	B	W	W	B	Left
#2	B	W	W	W	W	B	W	B	B	W	Right
#3	B	B	W	W	B	W	B	W	W	B	Right
#4	W	W	W	B	B	B	B	W	W	W	Left
#5	W	B	W	B	W	B	W	B	W	B	Right
#6	B	B	B	W	W	B	W	W	W	B	Left
#7	W	B	W	W	B	W	W	W	B	W	Right
#8	B	W	B	W	W	W	W	B	B	W	Left
#9	W	B	W	W	W	B	B	W	B	W	Left
#10	W	B	B	W	W	B	B	B	W	B	Right
#11	W	B	W	B	B	B	B	W	B	W	Left

The hand-raising decision was made as follows. Prisoner #3 would record the colors of the dots on the foreheads of the other prisoners in cyclic order, as shown in the following chart. Since the number of Ws along the main diagonal, shown in boldface, was even, prisoner #3 should raise his right hand.

Prisoner	Colors of Dots									
#4	**W**	W	W	B	B	B	B	(W)	W	W
#5	W	**B**	W	B	W	B	W	B	(W)	B
#6	B	B	**B**	W	W	B	W	W	W	(B)
#7	W	B	W	**W**	B	W	W	W	B	W
#8	(B)	W	B	W	**W**	W	W	B	B	W
#9	W	(B)	W	W	W	**B**	B	W	B	W
#10	W	B	(B)	W	W	B	**B**	B	W	B
#11	W	B	W	(B)	B	B	B	**W**	B	W
#1	W	B	B	W	(B)	B	B	W	**W**	B
#2	B	W	W	W	W	(B)	W	B	B	**W**

To determine the color of the dots on his own forehead, prisoner #3 observed the hand-raising of the other prisoners. For instance, prisoner #7 raised his right hand. This action was determined by the color of 1 dot on the forehead of each of the prisoners other than #7 himself. These were shown in brackets in the chart above, along with the dot 7 on the forehead of prisoner #3. Of the 9 dots visible to prisoner #3, there were 2 white ones, an even number. Since prisoner #7 raised his right hand, dot 7 on the forehead of prisoner #3 must be black. Similarly, the colors of dots 1, 2, 3, 4, 5, 6, 7, 8, 9 and 10 on the forehead of prisoner #3 would be determined by the hand-raising of prisoners #2, #1, #11, #10, #9, #8, #7, #6, #5 and #4 respectively.

Chapter 33

The Selection of the Argonauts

After handling so many funerals on behalf of the Olympians, Heracles finally came across one in which he had a personal interest. In fact, he gave the eulogy himself.

"Jason and I were boyhood friends. It was prophesied that his uncle Pelias, King of Iolcus, would be killed by a young relative. Pelias invited all his cousins and nephews to a banquet and massacred them. That very evening, Jason was born. His mother had arranged for him to be smuggled away to Mount Pelion before Pelias came to kill him. As Jason grew up, he and I became fellow students under Cheiron, king of the centaurs. His major accomplishments include leading the Black Sea Adventure and taking part in the hunting of the Calydonian boar. He was mostly remembered as Jason the Argonaut, or Jason of the Golden Fleece."

"How did Jason die?" I asked Heracles when we were back at the gate-house.

"He was a former hero who had come down in the world. He was sitting in the shade of his ship, weeping and thinking of his past glories, when the prow of the ship fell on him."

"What had he been doing since his glory days?"

"In his old age, he just engaged in some low-level piracy. He stopped merchant ships from passing through the narrow Strait of Corinth, and took the merchants on board his ship. He would tell his crew to line them up and *fleece* them for *gold*."

"He must be very annoyed to be remembered as Jason of the Golden Fleece."

"That was not what it meant," Heracles gave a rare chuckle. "There really was a gold fleece. Queen Ino of Thebes wanted her stepson Phrixus out of the way, and persuaded her husband King Athamas to sacrifice him to my father. I happened to be passing by during one of my labors. I told King Athamas in no uncertain terms that my father loathed human sacrifices. At the same time, my father sent a winged golden ram down from Olympus, and it carried Phrixus and his younger sister Helle away. Unfortunately, Helle fell and drowned in the strait since called the Hellespont. At Colchis, Phrixus sacrificed the ram to my father, and hung its golden fleece in a temple, guarded by a huge serpent. He then married a daughter of King Aeëtes and had four sons."

"How did Jason and the golden fleece come together, and why is he known as Jason the Argonaut?"

"After he finished his studies, Jason returned to Iolcus. Pelias wanted him out of the way, but dared not kill him openly. So he told him to lead the Black Sea Expedition, to retrieve the golden fleece from Colchis. Pelias built a ship, the *Argo*, which could carry 50 heroes. Jason sent out a call for heroes all over Greece to join him. Many came to Iolcus, including me. I had finished my studies before him, and had already made a name for myself. The others wanted me to lead the expedition. However, I pointed out that I was still a slave of Eurystheus, and in any case, that honor must go to Jason."

"Did he get enough heroes?" I asked.

"More than enough," replied Heracles. "Including himself, there were 75 heroes altogether. Not wanting to offend anyone by turning them away, Jason persuaded Pelias to build another ship, the *Ogra*, which could carry 25 heroes, for a Red Sea Expedition."

"Would that work? Anyone would know that being put on the Red Sea Expedition was the same as being rejected. In any case, what happened to this group of heroes?"

"Apparently, they did reach the north coast of Africa. However, there was no water passage, and they had to carry the boat over land across the Suez Isthmus. They had left tracks which could become the route of a canal. They were attacked by a monster which ate the boat, and has since been known as the *Ogre*. It ate most of the heroes too. The survivors fled east and sought protection under the Persian King of Kings, Darius the Great."

"What happened to your group, then?"

"Jason got the 50 heroes he wanted on the *Argo*, and they have since been known as the Argonauts. My own tenure as an Argonaut was very short. I had with me an orphan named Hylas, acting as a cabin-boy. A naiad had fallen in love with the pretty boy and pulled him under the water. Not knowing this at the time, I wanted the Argonauts to declare war on the local peasants, as I thought they had stolen Hylas. Instead, the Argonauts left me behind and sailed on."

"So you don't know what actually happened in this Black Sea Expedition."

"Not until Jason told me afterwards. On the way, the *Argo* met Phineus, the eldest son of Agenor. He advised Jason to pray to Aphrodite when he got to Colchis. Then the Argonauts passed an island and found the birds from Lake Stymphalia. They had settled there after I drove them off. Later they rescued the 4 sons of Phrixus whose ship had been wrecked. They joined the Argonauts to replace me and 3 other members who had been lost in various accidents."

At Colchis, Jason prayed to Aphrodite. She sent Eros to shoot an arrow at Medea, the youngest daughter of King Aeëtes, the same witch who later married King Aegeus, Theseus's father. She fell madly in love with Jason and helped him steal the Golden Fleece by immobilizing the guarding serpent with a drug. She sailed away with the Argonauts.

Aeëtes's fleet pursued the *Argo* all the way to the island of Drepane halfway up the Adriatic Sea, and asked the King to surrender Medea and the golden fleece. The King said that if Medea had not yet married Jason, she must go home to her father with the golden fleece. However, if they were already married, then Jason would keep her, and count the golden fleece as his wedding present.

Medea knew about this, and married Jason at midnight. The next morning, the King told the Admiral that Medea and the golden fleece were both Jason's now. The Admiral dared not fight the Argonauts, but neither did he dare go back empty-handed. He asked for asylum instead, and this was granted him. Some months later, the news reached Aeëtes at Colchis. He died of rage.

When the Argonauts returned to Iolcus, Medea knew by magic that Pelias had ordered that Jason was to be murdered at sight. Medea disguised herself as an aged woman and visited Pelias, pretending to be a goddess who could make him young again. She cut up an old ram, and by magic, it came back to life as a young one.

Pelias was not convinced. Medea told him to close his eyes. She made cutting noises and cried out as though in pain, but was actually taking off her disguise. When Pelias opened his eyes again, he found a beautiful girl before his eyes. So he got his eldest daughter to cut him up, and he died. She was the young relative the prophecy had warned him about.

Jason hung up the golden fleece in a temple of Zeus. Then he sailed to Corinth. King Corinthus died suddenly. The Corinthians chose Jason as the new king. Then they found out that Corinthus had been poisoned by Medea. To remain their king, Jason agreed to marry someone else. Aphrodite cursed Jason for breaking his marriage vow. He lost his throne, and wandered miserably all over Greece.

"It had a sad end," I said, "but it was still an exciting story."

"He returned to Corinth in his old age and became a pirate, as I have told you."

"How did he manage to get around the difficulty of selecting the Argonauts? You were there then."

"Jason divided the heroes into three groups, with 25 in each. Two of the groups were the ones he wanted. He and Pelias would play a game."

"What was the game?"

"Six dots were placed on a circle, alternating in black and white. Pelias and Jason each chose a dot in his turn, with Pelias going first. For each player, the first dot could be chosen arbitrarily. Subsequently, each could only choose dots adjacent to dots already chosen by him. Whichever player got more black dots won the game. If Jason won, he got to choose two groups to go on the *Argo*. If Jason lost, someone else would choose two groups to go on the *Argo*, and Jason would take the third group to go on the *Ogra*."

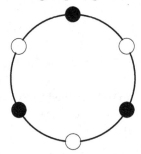

"How could Jason win? Pelias was sure to choose a black dot right away, and should be able to get one of the remaining two black dots."

"Think about it."

Solution

Because of the symmetry between the black and white dots, Pelias had only two distinct opening moves, taking a black dot or taking a white dot. The former seemed the better choice, since if Jason also took a black dot, the King would win the race to the last black dot, as I had thought. However, Jason had a surprise response, by taking the white dot diametrically opposite to the black dot taken by Pelias. He could now beat Pelias to each of the remaining two black dots.

What would happen if Pelias started by taking a white dot? Jason would take a black dot, not the one diametrically opposite but an adjacent one. Now each player moved along a semicircle, but Jason's semicircle contained two black dots. Paradoxically, although he chose second, Jason had a sure win.

Chapter 34

The Beef Feast

The Mount Olympus Cemetery was thriving. Heracles was summoned by Poseidon to go with him for the double funeral of his twin sons, Idas and Lynceus. As usual, I tagged along.

"Idas and Lynceus were mortal sons of mine," Poseidon said in the eulogy. "Idas was very skilful in throwing the javelin. Lynceus had outstanding eyesight. He was the lookout man on the *Argo*, and the first to spot the Calydonian boar. They had a beef-eating contest with the twins Castor and Polydeuces, sons of my brother Zeus. They had a disagreement which led to a fight. Polydeuces was the only survivor. Zeus brought Castor back from the dead. When I asked for the same treatment for my sons, he refused, claiming that his sons had won the fight."

Apollo was also in attendance. Afterwards, he came back with Heracles and me to the gatehouse.

"At one point," said Apollo, "Idas and I chased after the same girl named Marpessa. He had the audacity to fight me for her. My father liked Idas for his courage, and allowed Marpessa to choose between us."

"Whom did Marpessa choose?" I asked.

"Marpessa chose Idas. She said that gods usually abandon their mortal wives, but Idas would love her life."

"From what I have seen," said Hebe, "she was absolutely right."

"Do you know anything about this beef-eating contest?" Heracles asked.

"As a matter of fact, I do," replied Apollo. "The four of them decided to have some fun, and stole 101 cows from Iasus, King of Arcadia. They could not decide how to divide the spoil, because 101 cows would not come out in 4 equal portions."

"Did they fight over the cows then?" I asked. "However, it couldn't very well be the case, or it would not be called a beef-eating contest."

"No," said Apollo. "Idas killed 1 cow, cut it up into 4 equal shares and gave one to each of them. If they had any sense, they would eat their shares, and divide the remaining 100 cows equally among themselves. However, Idas had other ideas. He promised that whoever finished eating first would get 50 cows, as well as whoever finished second."

"So that was the beef-eating contest," said Hebe. "That seems fair."

"That was what the crafty Idas wanted everyone to think. He had already sharpened his knife for cutting beef. He finished before any of the other three had eaten more than half of their shares. Then he helped Lynceus to finish second by eating some of Lynceus's beef for him."

"Idas cheated," I complained, "and Lynceus ate only a part of his own share. Neither really deserved a prize."

"Outraged, Castor and Polydeuces hid in a hollow oak tree, preparing to murder their cousins. Lynceus saw them, and told Idas to hurl a javelin at the oak tree, killing Castor. Polydeuces rushed out and ran a spear through Lynceus. As Idas examined Lynceus, Polydeuces killed him too. After my father brought Castor back to life, he told them to take the 100 cows to the Mount Olympus Inn, for a beef feast in memory of their dead cousins."

"Does that mean I have to organize this feast?" asked Heracles.

"What do you expect?" laughed Apollo. "Castor and Polydeuces have given specific instructions. The 100 cows are all of roughly the same size. The beef must be divided equally among the invited families. Parts of the same cow may only be served to either 1 family or 2 families."

"How many families are invited?" I asked.

"Castor and Polydeuces are leaving that to Heracles. Good luck."

After Apollo had left, Heracles said to Hebe and me, "I am going to take a nap because there will be much for me to do. Meanwhile, you two have to figure out how many families I should invite."

"We cannot invite more than 200 families," I pointed out. "Otherwise, some cow will be served to more than 2 families. We can invite exactly 200 families, serving each half a cow."

"Knowing the citizens from around this area," said Hebe, "every family invited will be licking its collective lips. So Heracles can go ahead and book 200 tables at the inn."

"I wonder what all the possible values of the number of invited families are," I wondered, "if we are to obey the conditions laid down by Castor and Polydeuces."

"We can certainly invite any number of families which is a divisor of 100, namely 1, 2, 4, 5, 10, 20, 25, 50 and 100 itself. Can you think of any other?"

Solution

We already knew some of the possible values. The remaining values fell into two cases. In the first case, the number of families was less than 100. Then each fair share was more than 1 whole cow. In the second case, the number of families was more than 100. Then each fair share was less than 1 whole cow.

Could we invite 3 families? Then each fair share was 3 cows plus one-third of a cow. At first, it appeared that one cow would have to be shared among the three families. However, we could cut each of 2 cows into 2 pieces, the larger one consisting of two-thirds of a cow and the smaller one consisting of one-third of a cow. Then each of two families could have 33 cows and one small piece, while the third family had 32 cows and both large pieces. Thus 3 was a possible value.

We could not handle the potential values one at a time. Then I had an idea. I would have the families line up, and each took a fair share, starting from where the last family left off. Since each fair share was more one whole cow, no cow would be served to more than 2 families. This scheme worked for any number of families up to 100.

"I would love to see the families being fed this way," laughed Hebe. "The next potential value is 101. Consider each cow as divided into 101 units. Then each fair share consists of 100 units. The method you have just described also works here."

"If the number of families is 102," I said, "consider each cow as divided into 51 units. Then each fair share consists of 50 units. The same method still works. This suggests that other possible values are 100 plus a divisor of 100, as summarized in the following chart."

Families	101	102	104	105	110	120	125	150	200
Units/Cow	101	51	26	21	11	6	5	3	2
Units/Share	100	50	25	20	10	5	4	2	1

"I see," said Hebe. "For each possible value, the number of units in each fair share is always one less than the number of units in each cow. Hence there is no possibility that the same cow will be shared among more than 2 families. However, you must still prove that there are no other possible values."

"With more than 100 families," I said, "each cow must be shared between exactly 2 families. This suggests a graphical representation. In the diagram below, a dot stands for a family and a line stands for a cow. A dot lies on a line if and only if the family represented by the dot is being served a part of the cow represented by the line."

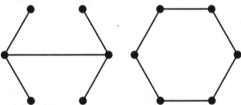

"Good idea," Hebe nodded. "The overall picture falls into a number of connected pieces. I claim that if I take away any of the lines in a piece, it will be split into two connected parts."

"That is true in the diagram above on the left, but not so in the diagram above on the right."

"This because there is a closed path or cycle," said Hebe. "However, in a cycle, the number of lines is equal to the number of dots. This means that those families are sharing at least as many cows, which puts each fair share to be at least 1 cow. Since this cannot be, my claim is justified."

"You still have to prove that all the pieces are of the same size."

"When all the lines have been removed from a piece, each dot by itself becomes a connected part. Since the removal of each line increases the number of parts by 1, the number of dots in a piece must be exactly one more than the number of lines. In a piece with k lines and $k + 1$ dots, each fair share is $\frac{k}{k+1}$ of a cow, which is different for different values of k. It follows that all the pieces are of the same size, and the only possible values greater than one hundred are just those listed in the earlier chart."

Chapter 35

The Hunters' Competitions

"In the funerals of Jason, Idas and Lynceus, sir," I asked Heracles, "the hunting of the Calydonian boar was mentioned. What can you tell me about this event?"

"All I know, Eubie," said Heracles, "is that it had happened some time ago. I need you to run a few errands for me, as I have to prepare for a visitor. He was in the hunt himself, and can tell you all about it."

When I returned to the gatehouse, I saw not one but two Heracles.

"Am I seeing double, sir?" I asked.

"You are not, Eubie," laughed Heracles. "Otherwise, you would be seeing four of us. This is my twin brother Iphicles."

"Pleased to meet you, sir," I said.

"My brother has told me a lot about you, Eubie. My life has been far less eventful than his. The only exciting thing I have ever done was to take part in the hunting of the Calydonian boar. I understand that you want to hear all about it."

"Yes, very much!"

"It happened just before the First Siege of Thebes. There was a huge wild boar in Calydon which killed many farmers and trampled many cornfields. Prince Meleager sent out heralds, inviting heroes from all over Greece to come and hunt it. Whoever killed the boar would be allowed to keep the skin."

"How many heroes showed up?" I asked.

"There were 16 of us. First of all, there was Prince Meleager of Calydon, arguably the best spearman in Greece. Several of the other heroes had been

Argonauts — such as Jason himself; Amphiaraus of Argos, who would soon die in the First Siege of Thebes; Peleus, King of the Myrmidons; and Ancaeus, the *Argo's* steersman. Theseus was there, along with Princess Atalanta of Arcadia. She was a famous huntress and the fastest runner in the world. There were the twins Castor and Polydeuces, the twins Idas and Lynceus, and 4 other youths."

"How did the other heroes take to Atalanta?" Hebe asked.

"She was fine with us except for Ancaeus, who refused to hunt with a woman! Meleager put his foot down and confirmed Atalanta's inclusion. He told Ancaeus to leave if this was not acceptable to him. Ancaeus did not want to go away because wanted the boar skin. To show that Atalanta had a rightful place in the hunt, Meleager ordered a trial, to obtain a complete ranking of the hunters according to their various skills."

"How was the trial organized?" asked Heracles.

"The 16 of us would compete one-on-one in several rounds. Each competition would determine which of the two hunters was the stronger."

"A complete ranking would take days," I exclaimed.

"Indeed, it took 10 days, because each hunter was allowed to participate in only one competition per day. Jason got the top ranking, followed by Theseus, Meleager and Atalanta. They formed the first group. They were followed by Amphiaraus, myself, Ancaeus and Peleus. We formed the second group. The two pairs of twins formed the third group and the other 4 youths formed the last group."

"In his eulogy," Heracles said, "Poseidon mentioned that Lynceus was the first to spot the boar."

"Indeed he was. He blew the horns. The boar rushed out and killed 3 of the youths. The fourth one escaped by climbing a tree. The twins threw javelins at the boar. All 4 missed."

"What did you do?" I asked.

"I threw javelins too. Only one found the target, but it just scratched its shoulder. Ancaeus tripped and fell. Atalanta shot the side of the boar's head. The wounded boar charged and tore Ancaeus to pieces. Peleus threw his javelin wildly. It bounced off a tree and killed him."

"So 5hunters had died already," I said, keeping count.

"At last Amphiaraus shot the right eye of the boar. It went for him, but fortunately Meleager drove his spear into the boar's heart. The monster fell dead. Meleager awarded the skin to Atalanta, stating that her arrow would have caused the boar's death in any case."

"Wow! That was quite an event," said Hebe. Turning to me, she added, "We had better leave the boys alone, because they must have a lot to catch up on. Let us see how '6 hunters could be completely ranked in ten days."

Solution

We went outside and sat under a tree.

"The number of participants in such competitions are usually a power of 2, and 16 is indeed one of them," I said. "Let us start with only 1 hunters. The ranking takes just 1 day. With 4 hunters, I can form two ranked pairs (A_1, A_2) and (B_1, B_2) in 1 day. We need to merge them into a ranked quartet."

"The usual way," said Hebe, "is to have A_1 go against B_1 to see who is first, and A_2 against B_2 to see who is last. These can take place on day 2. The ranking can be completed in at most 1 more day."

"With 8 hunters, we can form two ranked quartets (A_1, A_2, A_3, A_4) and (B_1, B_2, B_3, B_4) in 3 days," I said. "We need to merge them into a ranked octet. On day 4, we can have A_1 go against B_1, A_2 against B_2, A_3 against B_3 and A_4 against B_4. However, it is not clear how to proceed from here."

After a while, Hebe said, "Let us do this in an unusual way. Let A_1 go against B_4, A_2 against B_3, A_3 against B_2 and A_4 against B_1 on day 4. If either A_4 or B_4 wins, the ranking is complete."

"So we may as well assume that both A_1 and B_1 win," I said. "Suppose A_3 wins. Then the top four hunters are A_1, A_2, A_3 and B_1. On day 5, B_1 go against A_2. On day 6, if B_1 has won the day before, B_1 will go against A_1 for first place. If B_1 has lost the day before, B_1 will go against A_3 for third place. The bottom 4 hunters can be ranked in a similar way, with A_4 going against B_3 on day 5."

"The case where B_3 wins over A_2 is analogous," Hebe said. "Henceforth, we may assume that all of A_1, A_2, B_1 and B_2 win on day 4. Then they are the top 4 hunters. We can obtain two ranked quartet in 2 more days, either as before or by our unusual way, completing the ranking."

"So in 6 days," I said, "the 16 hunters can be sorted into two ranked octets $(A_1, A_2, A_3, A_4, A_5, A_6, A_7, A_8)$ and $(B_1, B_2, B_3, B_4, B_5, B_6, B_7, B_8)$. On day 7, A_1 goes against B_8, A_2 against B_7, A_3 against B_6, A_4 against B_5, A_5 against B_4, A_6 against B_3, A_7 against B_2 and A_8 against B_1. If either A_8 or B_8 wins, the ranking is complete."

"So we may as well assume that both A_1 and B_1 win," Hebe said. "Suppose A_7 wins. Then the top 8 hunters are A_1, A_2, Aa_3, A_4, A_5, A_6, A_7 and B_1. On day 8, B_1 go against A_4. On the ninth day, if B_1 has won the day before,

B_1 will go against A_2. If B_1 has lost the day before, B_1 will go against A_6. What will happen on day 10 will be decided in a similar manner. The bottom 8 hunters can be ranked in a similar way, with A_8 going against B_5 on day 9."

"The case where B_7 wins over A_2 is analogous," I said. "Henceforth, we may assume that all of A_1, A_2, B_1 and B_2 win on day 7. Suppose A_6 wins over B_3. Then the top 8 hunters are A_1, A_2, A_3, A_4, A_5, A_6, B_1 and B_2. On day 8, B_1 goes against A_4 while B_2 goes against A_3. If A_4 wins, we can merge (A_5,A_6) with (B_1,B_2) in 2 more days. If A_4 loses but A_3 wins, then the top 4 hunters are A_1, A_2, A_3 and B_1 and the ranking among the top 8 can be completed in 2 more days as before. Finally, if A_3 also loses, we can merge (A_1,A_2) with (B_1,B_2) in 2 more days. The bottom 8 hunters can be ranked in a similar way."

"The case where B_6 wins over A_3 is analogous," said Hebe. "Henceforth, we may assume that all of A_1, A_2, A_3, B_1, B_2 and B_3 win. Suppose A_5 wins over B_4. Then the top 8 hunters are A_1, A_2, A_3, A_4, A_5, B_1, B_2 and B_3. On day 8, B_1 goes against A_4, B_2 goes against A_3 while B_3 goes against A_2. If A_4 wins, then the ranking among the top 8 can be completed in 2 more days as before. If A_4 loses but A_3 wins, then the ranking among the top 4 can be completed as before, while (A_4,A_5) can be merged with (B_2,B_3), all in 2 more days. If A_3 loses but A_2 wins, (A_1,A_2) can be merged with (B_1,B_2) while the ranking among the next 4 can be completed as before, all in 2 more days. Finally, if A_2 also loses, then the ranking among the top 4 can be completed in 1 more days as before. The bottom 8 hunters can be ranked in a similar way."

"The case where B_5 wins over A_4 is analogous," I said. "The only remaining case is where all of A_1, A_2, A_3, A_4, B_1, B_2, B_3 and B_4 win. Then they are the top 8 hunters. We can merge the two ranked quartets within each octet in 3 more days. Thus the complete ranking would indeed take 10 days. What a remarkable problem this turns out to be!"

Chapter 36

The First Rebellion of the Giants

One day, I was told that someone asked to meet me in the Mount Olympus Inn. I hurried downhill, and found Silenus in the lobby.

"What brings you here, sir?" I asked.

"I was just roaming about, Eubie," he said. "However, tumultuous times are upon the Olympians. Have you heard of Gaia?"

"I believe she is commonly known as Mother Earth, sir," I said.

"Yes, she is the mother of Cronus and the grandmother of Zeus. She has three sons by her titan husband Uranus. Cronus is the eldest. The middle boy is Coeus. He has only one daughter Leto, who was seduced by Zeus. The twins Apollo and Artemis were the fruits of that affair. The youngest son is Iapetus, and he in turn has three sons. The eldest is Atlas whom Perseus turned into Mount Atlas. His daughter Maia was seduced by Zeus and gave birth to Hermes, as I have told you before. The middle boy Prometheus is unmarried. The youngest is Epimetheus. His daughter Dione was seduced by Zeus and gave birth to Aphrodite."

"There are also Minos by Europa, Heracles by Alcmena, as well as Castor and Polydeuces by Leda," I commented. "I am losing count."

"That is why Zeus is often described as *omnipotent*," Silenus chuckled. "Coming back to Mother Earth, she has other descendants besides titans. They include the cyclopes and the giants."

"Why does she want to cause trouble?"

"The current fuss was triggered by our friend Heracles. He was challenged to a wrestling match by Antaeus, a giant son of Mother Earth. Every time he threw Antaeus to the ground, Antaeus rose again stronger than ever as his strength was renewed by contact with Mother Earth. So Heracles lifted Antaeus off the ground and crushed him to death. In revenge, Mother Earth went to Thrace and raised an army of enormous giants. They are coming to attack the Olympians. I think you should go back to Eleusis, and I will come with you."

"I don't like the idea of running away in face of trouble. Besides, I cannot possibly be any safer than being with Heracles."

"That is true," conceded Silenus. "I now have no reason to go back to Eleusis immediately. I will continue my vacation elsewhere."

The Olympians had also heard the news. They decided to launch a pre-emptive strike, led by Heracles. I accompanied them as they marched to Thrace.

Heracles shot Alcyoneus, the leader of the giants, but could not lift him off the ground. So he dragged Alcyoneus over the Greek frontier into Scythia and started clubbing him to death. Meanwhile, the other giants made a charge. Without Heracles, the Olympians could not hold the line. They beat a hasty retreat.

I was worried about Hebe, and was glad to find her hiding with Hestia and Persephone in a linen closet in the gatehouse. The best I could do was to stay with them.

The giants besieged the Mount Olympus Palace, beating back several futile sorties. Eventually, the gate was breached. A giant named Porphyrion caught Hera and tried to strangle her. Eros shot him in the heart with his little bow, so that he fell madly in love, and gave Hera great slobbering kisses. Angrily, Zeus threw a thunderbolt at him, but it was deflected by his shield. He kissed Hera again and again.

Heracles returned just in time to break the giant's neck and hold him aloft until he died. Encouraged by the return of their champion, the Olympians threw themselves at the giants. The resulting melee was heated and confusing. The conflict ended only when the three Fates were aroused and charged out.

All the giants fled, for nobody could fight against the Fates. I went with the Olympians in their pursuit of the retreating enemy. Half-heartedly, the giants made their last stand in Arcadia. Ares and Athena fought well with their spears and shields, ably assisted by Zeus with his thunderbolts and Poseidon with his trident. Hermes had borrowed Hades's helmet of invisibility and stabbed the enemy from behind. Artemis, Dionysus and Apollo shot at the giants from safe distances.

The giants were overwhelmed and annihilated. Upon the triumphant return of the Olympians, Hera came up and thanked Heracles for getting rid of that disgusting Porphyrion.

She said, "I am ashamed to think how badly I have treated you before."

"Think nothing of it, Your Majesty," said Heracles graciously.

Apparently, the Fates had been keeping records of the battle. They presented a list of the combatants in non-ascending order of the numbers of giants they had killed. Naturally, Heracles was at the top of the list. He was followed by Ares, Athena, Zeus, Poseidon, Hermes, Artemis, Dionysus, Apollo, Demeter, Hephaestus, Aphrodite and Hera in that order. The Fates also observed that for any group of 5 combatants, at least four-fifths of the giants killed by the group was killed by 1 of its members.

"That means Heracles had killed four-fifths of the giants," I said to Hebe after she had come out of the linen closet and sent Hestia and Persephone away.

"That is not what the Fates said," she remarked.

"I know, but doesn't my conclusion follow from their observation?"

"I don't think so. Of course, if only 5 of them kill any giants at all, then you are right. Consider the following example."

She drew up the following chart.

Combatant	Heracles	Ares	Athena
# of giants killed	380	76	16
Combatant	Zeus to Artemis	Dionysus to Hera	
# of giants killed	1 each	0 each	

"The total number of giants killed is $380 + 76 + 16 + 4 = 476$," I said as I computed, "and 380 is indeed less than four-fifths of it. However, do your data satisfy the Fates' observation?"

"We will check that of the giants killed by any group of 5 combatants, at least four-fifths are killed by 1 of the members. We can safely dismiss everyone from Dionysus to Hera. Any group of 5 combatants must include at least 2 of those from Zeus to Artemis. If it includes all 4 of them, then the fifth member has certainly killed at least four-fifths of the giants killed by the group."

"Suppose it includes 3 of them," I chipped in. "If the other 2 are Ares and Athena, then the group has killed 96 giants and Ares is credited with 76 of them. This is more than four-fifths. It is certainly true if either Ares or Athena is replaced by Heracles."

"Finally," concluded Hebe, "suppose the group includes only 2 of those from Dionysus and Artemis. Then the other three must be Heracles, Ares and Athena. The group has killed 474 giants and Heracles is credited with 380 of them. This is more than four-fifths."

"Can four-fifths be replaced by a smaller fraction, say three-quarters?"

"Most certainly, Heracles must have killed at least three-quarters of the giants. However, can you prove it?"

At that point, Heracles came in and told us that there was a victory banquet, and it was time to go. We did not returned to the gatehouse until just before midnight. Unable to sleep because of all the excitement, I sat down and tried to come up with the proof that Hebe wanted.

Solution

The first group of 5 combatants I considered consisted of Heracles, Ares, Athena, Zeus and Poseidon. Since Heracles had killed at least four-fifths of the giants killed by this group, the number of giants he killed is at least four times the total number of giants killed by the other 4.

With 13 not being a multiple of 5, I considered overlapping groups. The second consisted of Poseidon, Hermes, Artemis, Dionysus and Apollo, while the third consisted of Apollo, Demeter, Hephaestus, Aphrodite and Hera. As before, the numbers of giants killed by Poseidon in the second group and by Apollo in the third group were at least 4 times the total number of giants killed by the other 4.

Now I added up these inequalities, and found that the total number of giants killed by Heracles, Poseidon and Apollo was at least 4 times the total number of giants killed by the 12 combatants other than Heracles. Canceling once the contributions by Poseidon and Apollo from both sides, the number of giants killed by Heracles was much more than 3 times the total number of giants killed by the others.

Chapter 37

The Second Rebellion of the Giants

Despite the recent victory of the Olympians, all was not well. Heracles, who had done so much, was granted a vacation. While he was away, Mother Earth raised another army of giants to invade Mount Olympus. They occupied Mount Ossa, which was nearby, in order to throw rocks from its top at the palace.

The Olympians sent me to find out more. They assumed that as a small boy, I would not be noticed by the giants. From close range, I observed that the side of Mount Ossa had 4 ledges at equal vertical distance apart. In each move, 2 giants might climb together onto ledge #1. Alternatively, if 2 giants were on the same ledge, 1 of them could get on the shoulders of the other and climb onto the next ledge above, while the supporting giant would roll down to the next ledge below, or to the ground from ledge #1.

To avoid shaking up the foundation of the mountain, the moves must be made 1 at a time, and only 5 moves could be made in 1 day. Still, by the morning of day 3, a giant had reached ledge #4 at the top of Mount Ossa. I recorded their progress in the diagram below.

Moves	1	2	3	4	5	6	7	8	9	10	11
4th Ledge											●
3rd Ledge					●	●	●	●	●	●●	
2nd Ledge		●	●	●●			●	●	●●		●
1st Ledge	●●		●●		●	●●●	●	●●●●	●	●●	●●

However, while this plan sounded great, it was not effective since Mount Olympus was taller than Mount Ossa. The operation was halted and no more giants were sent up the ledges. The giants already there remained in place in case the plan were reactivated.

Meanwhile, the Olympians had received a declaration of war from the Aloeids twins, the leaders of the giants.

"This is an outrage," Zeus thundered. "Ephialtes, the elder twin, demands Hera as his bride, and Otus, the younger twin, claims Artemis for himself."

"What can we do?" asked Hera. "It has been prophesied that no gods and no other mortals could kill either of them. We are doomed even if Heracles returns in time. Although he will not lose, he cannot win either."

"To compound our problem," added Hephaestus, "I just observed that Mount Pelion, has somehow been moved to the top of Mount Ossa. Only Mother Earth could pull off such a trick."

"There are 5 ledges on the side of Mount Pelion." said Hermes. "They are at the same vertical distance apart as those on the side of Mount Ossa. Its ledge #1 is now ledge #5 on the side of the combined mountain. The giants have just resumed climbing the ledges. If a giant reaches ledge #9, it would be a disaster for us since the combined mountain is taller than Mount Olympus. Father, you have to knock Mount Pelion back to where it should be with your thunderbolts."

"The palace is under siege," said Zeus. "We have to dispose of the twins before I can go out and get close enough to throw my thunderbolts. This is not going to be easy without Heracles. Eubie, how many more days would it take for a giant to reach ledge #9?"

I replied, "I can't answer that question right away, Your Majesty. I will have to do some calculations. As a rough estimate, I would guess 20 days."

"That leaves us little time," said Hera. "Heracles may not come back in time. We have to figure out a way to deal with the twins."

"I have a plan," said Artemis. "I have written a reply to their demand."

"What are you saying?" asked Apollo.

Artemis read her reply aloud, "My dearest Otus, I would love to be married to you, even though my mother said that she would never marry your twin brother. Love and kisses, Artemis."

Later, a reply came from the twins. Offended, Ephialtes decided to claim Artemis for himself and asked her to choose between the twin brothers.

Artemis again read aloud her reply, "My dearest Ephialties and Otus, I am sending you a hind. Please release it and then try to hunt it down. I will marry whichever one of you is the better archer. Love and kisses, Artemis."

Then she disguised herself as a hind. From the rampart above the palace gate, we saw the twins let the hind loose in a wood and go after it. For days, Artemis darted around the trees. Finally, after 19 days, she got the opportunity she wanted. She rushed between the twins. They seized their bows and shot simultaneously at her from opposite sides. Artemis ducked out of harm's way. The twins were killed, not by gods or other mortals, but by each other.

The Olympians now charged out with their weapons. With their leaders dead, the giants were demoralized and fled. Zeus was able to throw a few thunderbolts and knocked Mount Pelion off the top of Mount Ossa.

After another victory banquet in the palace, I returned to the gatehouse, but not as late as on the previous occasion. Then I remembered that I had not even thought about the question raised by Zeus.

Solution

I explained it to Hebe, and we worked on it together.

"In the excellent diagram you have drawn," she asked, "what can you observe about move 11?"

"Move 11 was the first time a giant reached ledge #4," I said. "Ledge #3 was vacant because 2 giants were there in move 10."

"Why couldn't there be more than 2 of them?"

"Then the giants could have reached ledge #4 sooner. The other giant on ledge #3 in move 10 had rolled down to ledge #2. He would be the only one for the same reason as before. However, why were there 2 giants on ledge #1? Shouldn't there be only 1?"

"There is a subtle reason. Call the total number of giants on the odd-numbered ledges the magic number. I claim that it must be even. This is true initially because there are no giants on any ledge. A move from an odd-numbered ledge decreases the magic number by 2. A move from an even-numbered ledge increases the magic number by 2. This is also true if we regard the ground as the zeroth ledge. Thus my claim is justified."

"That is neat!" I exclaimed. "After the move which gets a giant on ledge #5 for the first time, ledge #4 must be vacant. There must be a giant on each of ledges #3 and #2, and 2 on ledge #1. Let me expand my diagram and see if that is indeed the case."

Moves	11	12	13	14	15	16	17	18	19	20
5th Ledge										•
4th Ledge	•	•	•	•	•	•	•	•	••	
3rd Ledge			•	•	•	•	•	••		•
2nd Ledge	•	••			•	•	••		•	•
1st Ledge	••		•	•••	•	•••	•	••	••	••

"You can see that I am right. After getting a giant on ledge #4 on move 11, it took another 9 moves to get one on ledge #5. Could we have computed this number without drawing the expanded diagram?"

"Let me see. Our intermediate objective is to get a giant in each lower ledge, except possibly the first one. We already have 2 giants on ledge #1, and 1 on ledge #2 after move 5. It takes 2 moves to fill the vacancy on ledge #3."

"However, this creates a vacancy in ledge #2," I observed.

"Moreover, there is now only 1 giant on ledge #1. We must use an extra move to bring in reinforcement from the ground. Now it takes only 1 more move to fill the vacancy on ledge #2."

"So this stage can be completed in $2 + 1 + 1 = 4$ moves. What next?"

"We have to make another reinforcement move. Then it takes 4 more moves to get a giant on ledge #5. The total number of moves required is indeed $4 + (1 + 4) = 9$."

"Can you now compute how many more moves after move 20 before a giant can reach the ledge #6?"

"Well, the first stage takes $3 + 1 + 2 + 1 + 1 = 8$ moves and the second stage takes $1 + 5 = 6$ moves. A giant can reach ledge #6 in $8 + 6 = 14$ moves."

"Essentially correct," said Hebe, "but you make one error. When you try to fill the vacancy in ledge #4, you start with 2 giants on ledge #1. When you try to fill the vacancy in ledge #3, there is only 1 giant there. This is why you need the reinforcement move. However, when you try to fill the vacancy in ledge #2, there will be 2 giants on ledge #1 again. Thus the first stage takes only $3 + 1 + 2 + 1 = 7$ moves, and a giant can get to ledge #6 on move $20 + 13 = 33$. How many giants will be on ledge #1 then?"

"By your magic number argument, there will only be 1 of them. To get a giant on ledge #7, we must start with a reinforcement move. The first stage will take $1 + 4 + 3 + 1 + 2 + 1 = 12$ moves and the second stage will take $1 + 6 = 7$ moves. This means that a giant can reach ledge #7 on move

$33 + (12 + 7) = 52$. Again, there will only be 1 giant on ledge #1. To get a giant on ledge #8, we must again start with a reinforcement move. The first stage will take $1+5+4+1+3+2+1+1 = 18$ moves and the second stage will take $1 + 7 = 8$ moves. A giant can reach ledge #8 on move $52 + (18+8) = 78$. This time, it is back to 2 giants on ledge #1."

"Leave something for me to do," shouted Hebe, caught up in the excitement. "To get a giant on top of the combined mountain, the first stage will take $1 + 6 + 5 + 1 + 4 + 3 + 1 + 2 + 1 = 24$ moves and the second stage will take $1 + 8 = 9$ moves. This means that the giants can complete their mission on move $78 + (24 + 9) = 111$!"

"Since $111 - 11 = 100$ and $100 \div 5 = 20$, the rough estimate I gave Zeus turned out to be sharp! Artemis saved the Olympians just in time."

Chapter 38

The Third Rebellion of the Giants

Mount Olympus enjoyed a period of relative peace and quiet. Eros and I went to have some fun in the Mount Olympus Arcade.

"Things have changed a lot since you were the temporary God of Redemption," he informed me. "Instead of winning wreaths and redeeming them for vases, clients now win coins. However, after a client has made a net gain in a game, he may not play it again."

Eros found an easy game called the "Wild Goose Chase". The playing field was a straight path on which a wolf was crouching. Eros would purchase wild geese at 1 coin per piece, and the machine would place them along the path in locations of its choice. Eros's turn was first. He then took alternating turns with the machine.

During his turn, he could move the wolf 1 unit in either direction. During the machine's turn, it could move any 1 of the geese 1 unit in either direction. If the wolf caught a goose, Eros would win 5 coins.

Eros knew that if he bought only 1 goose, the machine would place it just out of reach of the wolf. He reasoned that if he bought 2 geese, the wolf would surely catch ' if it ran at them. However, he did not expect that the machine would place ' goose on each side of the wolf, again just out of its reach.

Having lost 2 coins, Eros gave up, but I urged him to persist. He bought 3 geese. This time, he was assured a victory, since no matter where the machine placed the geese, at least 2 of them had to be on the same side of the wolf. The wolf could just run at them. So Eros won back the 2 coins he had lost.

A game caught my eye. It was called the "Tower of Delphi". There were three pegs in the playing board and 3 disks of different sizes. These disks were stacked on the first peg in ascending order of size from the top, forming the Tower of Delphi. The objective was to transfer this tower to the third peg.

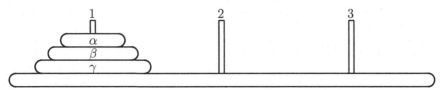

The rule was that Eros might only move a disk on top of a peg to the top of another peg, and might not place a disk on top of a smaller disk. Eros had to pay one coin for each move, and would win ten coins upon completion of the task.

Eros felt certain that he could win this game, but he lost a lot of money by making useless moves.

Finally, I showed him a 7-move solution: α to peg 3, β to peg 2, α to peg 2, γ to peg 3, α to peg 1, β to peg 3 and finally α to peg 3. He summarized it as $\alpha\beta\alpha\gamma\alpha\beta\alpha$.

Suddenly, we felt some tremor on the ground. Everybody rushed outside lest the building would collapse. From the talk of the people, it appeared that there was an earthquake in Thrace, and the aftershock was extending southward.

We passed several worrying days. Then a monstrous giant, the largest ever seen on earth, appeared in the horizon. He had a donkey's head, with ears that reached the sky and wings that blocked out the sun. He was moving slowly but steadily towards Mount Olympus, at the head of an army of giants.

It turned out that Mother Earth was causing trouble yet again. From the people fleeing before him, I learnt that this giant was Mother Earth's latest creation, and she named him Typhon.

All the Olympians were terrified. They asked Heracles to station himself outside the back gate and guard the north end of the palace. Typhon rushed towards Heracles, spouting flame. He was the toughest opponent Heracles had ever faced. While holding his own, Heracles was not gaining the upper hand. In time, he would be driven back inside the palace wall. The giants would then besiege the palace.

The Olympians decided that discretion was the better part of valor, and escaped through the main gate at the south end of the palace while they still had a chance. Their flight took them across the Mediterranean Sea to Egypt.

When told that his intended victims had escaped, Typhon disengaged from Heracles and led part of his army to pursue the Olympians all the way to Africa. Heracles stayed behind and organize the defense of the palace against the remaining giants. Since I traveled lighter and therefore faster than Typhon's horde, Heracles sent me to warn the Olympians that Typhon was on their trail.

The Olympians were stricken with *Typhon fever* and lost their heads. Zeus disguised himself as a ram, Hera as a cow, Poseidon as a horse, Demeter as a pig, Hephaestus as a quail, Ares as a boar, Aphrodite as a dove, Apollo as a crow, Artemis as a wildcat, Hermes as a crane and Dionysus as a tiger. Only Athene refused to disguise herself. When Typhon eventually made his appearance, she took him on bravely by herself. She was clearly no match for him. Fortunately, Typhon was more interested in chasing the Olympians in disguise.

Typhon made leaps of length 3 stadions at the Olympians, pausing for breath between leaps. During a leap, the petrified Olympians were rooted to their spots. During a pause, one and only one of the Olympians ran away, covering a distance of 3 stadions.

Although Typhon was not yet in range, the Olympians were running around like headless chickens. If they hung together, Typhon was sure to catch and kill some of them. I had to be quick. So I took charge immediately and gave them very simple instructions, for they were in no shape to understand complicated strategies. That kept them safe.

Then Athene came forth again. She called Zeus a coward, and said that she was ashamed to acknowledge herself his daughter. Zeus blushed, turned back into his proper shape, and flung a thunderbolt at Typhon from behind. This caught Typhon by surprise, and he was wounded in the shoulder.

With a yell of pain, Typhon seized Zeus, beat him black and blue, and then pulled out the sinews from his feet and hands to make him helpless. However, the pain in his shoulder was agonizing. He left Zeus in charge of a she-monster named Delphyne. He himself returned to the Mount Olympus Palace, to seek help from the Fates.

I cautiously followed the wounded Typhon. The Fates said nothing. They just handed Typhon some apples and went on spinning. He crunched the fruit in his huge teeth. However, the Fates had tricked him. The apples were poisoned. As the poison took effect, Typhon grew weaker and weaker. He sat helplessly outside the palace.

Then I received a nessage from someone asking to meet me in the Mount Olympus Inn. I hurried downhill, and found in the lobby Silenus's father Pan. He still refused to go up to the palace.

"My boy," Pan said, "I have brought you an excellent piece of news. While Typhon was here, I joined the Olympians in Egypt. They had discarded their disguises. Hermes and Apollo took me with them to Delphyne's cave at night. I let out my usual horrifying shout. That made Delphyne jump nearly out of her skin. Hermes slipped in unnoticed and stole Zeus's sinews from a chest under her bed. The theft was soon discovered, and she gave chase. Apollo lay in ambush and shot Delphyne dead. The Olympians are on their way back to Mount Olympus. Zeus wants the Fates to fix his sinews to his hands and feet again."

It did not take Zeus long to return to normal. He pelted the feeble Typhon with thunderbolts. Typhon retreated, pursued by a vengeful Zeus. I tagged along with Zeus. We caught up with Typhon in Sicily and Zeus hurled an enormous rock on top of him. It is now known as Mount Etna. From time to time, Typhon's fiery breath bursts into the air, spraying smoke and lava through the crater.

When everything had quieted down, Hebe asked me, "What simple instructions did you give the disguised Olympians?"

"Let me explain," I said, and told her that my idea was inspired by the game "Wild Goose Chase".

Solution

The key was to keep the disguised Olympians apart from one another, so that Typhon could only threaten one of them at a time. In a one-on-one race, Typhon could not win. To keep my instructions simple, the Olympians were asked to run along straight paths. As long as these paths did not cross and were far enough apart, Typhon could not threaten rwo or more Olympians at a time.

What I needed were straight paths which would not cross. So I told each Olympian to run along one of a family of parallel paths, 7 stadions apart from one another. Only the Olympian threatened by Typhon had to run. Thus Typhon could not catch any of them.

Chapter 39

The Ungrateful Refugees

Zeus was not in a good mood. His sinews were still giving him pain from time to time. He decided to stretch his limbs by visiting Arcadia, accompanied by Hermes, to check on the behavior of the inhabitants. To do so, they disguised themselves as ordinary mortals. They decided to take me along, because having a small boy in tow made two grown men less suspicious. Being already an ordinary mortal, I required no disguise.

I was shocked at how selfish and hard-hearted the people were, with no pity for the poor, nor sorrow for the homeless. These wicked people taught their children to be no better than themselves, and used to clap their hands, to urge them on, when they saw the little boys and girls run after some poor stranger, shouting at his heels, and hitting him with stones.

The Arcadians kept large and fierce dogs, and whenever a traveler appeared, these vicious animals rushed to meet him, barking, snarling and showing their teeth. Then they would seize him by the leg, or by his clothes, just as it happened; and if he were ragged when he came, he was generally a piteous object by the time he escaped.

Yet, when rich persons came in their carriages, or riding on beautiful horses, with their liveried servants waiting on them, nobody could be more civil and slavish than the Arcadians. They would take off their hats and make the humblest bows. If the children were rude, their parents would box their ears; and if a dog dared to bark, its master beat it with a stick, and tied it up without any supper.

"There is no hope for such people," thundered Zeus. "It is better to wipe them all out with an enormous flood."

"Let us check on one more family, father," pleaded Hermes.

We climbed a hill, and found an old couple at their cottage-door enjoying the calm and beautiful sunset. They seemed quite poor, and probably had to work very hard for a living. Yet they welcomed us with such friendly faces that there was no need of saying anything.

Hermes said, "This is quite another greeting than we have met elsewhere. Why do you live among such bad people?"

"The gods and goddesses put us here in order that we may make up for the unkindness of our neighbors."

"Well said!" cried Zeus, laughing. "Tell us all about yourselves."

"My name is Philemon," said the old man. "I toil hard in the garden. My wife Baucis is always busy with her spinning, or making a little butter and cheese with our cow's milk, or doing one thing or another about the cottage."

Baucis added, "Our food is seldom anything but bread, milk, and vegetables, with sometimes a little honey from our bee-hive, and now and then a bunch of grapes from our orchard. You are most welcome to share our humble meal."

"You are a good old couple," said Zeus. Forgetting himself, he added, "I will grant you one wish."

"We do not wish for anything," said Philemon. "We already have each other."

"Although we were not born on the same day," said Baucis, "we wish to die on the same day."

"Wish granted!" Zeus said, and threw a thunderbolt at them.

"What did you do *that* for?" cried Hermes, aghast.

"Oops! I wasn't thinking," said Zeus. Turning to the two piles of ashes, he added, "Sorry."

"A lot of good *that* does!" Hermes muttered.

Zeus waved his arms about. Instantly, an oak tree grew out of Philemon's ashes and a linden tree out of Baucis's ashes. Their boughs were twined together and wrapped round one another, so that each tree seemed to live in the other tree's bosom, much more than in its own. Hermes built a seat around both their trunks, under their kindly shades, welcoming any wayfarer who might pause there.

Having found no other worthy people, Zeus summoned a flood over Arcadia. It gradually spread to the rest of Greece. However, having killed Philemon and Baucis thoughtlessly, Zeus relented somewhat, and halted the torrential rain. He judged that he had given the survivors a sufficient warning to mend their ways.

King Deucalion and Queen Pyrrha of Thessaly were good people. An oracle had warned them of the coming flood. They built an ark and filled it with the king's flocks, herds and other possessions. After tossing about for nine days, their ark came to rest at the top of Mount Othrys. They sacrificed a ram to Zeus for sparing the mortals.

Many homeless refugees converged on Mount Olympus. The Mount Olympus Warehouses were cleared out and partitioned into apartments. Each family was housed in a separate apartment. Heracles was put in charge, and he appointed me his deputy.

Trouble began almost immediately. Now that their danger was over, the refugees complained about their accommodations. When Heracles told them that there were no alternatives, they began complaining that other families had better apartments.

"I have had it up to here with these trouble-makers, Eubie," Heracles said to me. "Tell them in no uncertain terms that I will only permit pairwise exchanges of apartments. Moreover, a family may participate in at most one exchange per day."

"Yes, sir," I said.

I had no idea what I was going to do, or could do. I prayed for a miracle, and I got it! I asked the refugees to submit a wish list, each family indicating a preferred apartment. I was amazed to discover that each family indicated a different one!

I still had to follow the rules Heracles had laid down. I began working on finding the minimum number of days in which all the refugees' wishes could be granted.

Solution

I went back to the gatehouse to get help from Hebe.

"The key observation, Eubie," she said, "is that the whole exchange can be broken down into a number of cyclic exchanges, with each family in only one cycle."

"What is a cycle or a cyclic exchange?" I asked.

"Suppose family 1 is happy with apartment 1. Then (1) is a 1-cycle. Suppose family 1 wants apartment 2 and family 2 wants apartment 1. Then (1,2) is a 2-cycle. The exchange within the cycle takes only 1 day. Since neither the number of families nor the number of apartments is unlimited, a cycle must be formed sooner or later."

I showed Hebe a hypothetical wish-list.

Family	1	2	3	4	5	6	7	8	9
Preference	1	3	5	6	7	4	2	8	9

"Am I right that the cycles are (1,9), (2,3,5,7), (4,6) and (8)?"

"Yes," she confirmed. "As long as each family prefers a different apartment, the wish list can always be broken down into cycles. If you can handle cycles, then you have solved the problem."

I first considered a 6-cycle and found a cyclic exchange that required only two days.

Apartments		1	2	3	4	5	6
Fami	Day 1	6	5	4	3	2	1
-lies	Day 2	6	1	2	3	4	5

On day 1, the families from opposite ends exchanged apartments, namely, family 1 with family 6, family 2 with family 5, and family 3 with family 4. On day 2, the family in apartment 1 was already in the right place. The remaining familes from opposite ends exchanged apartments, namely, family 1 with family 5, family 2 with family 4, while family 3 stayed put. Now every family was happy.

This method worked for all even cycles. I now considered an odd cycle, taking as an example a 7-cycle. Again the cyclic exchange required only two days.

Apartments		1	2	3	4	5	6	7
Fami	Day 1	7	6	5	4	3	2	1
-lies	Day 2	7	1	2	3	4	5	6

On day 1, the families from opposite ends exchanged apartments, namely. family 1 with family 7, family 2 with family 6, family 3 with family 5, while family 4 stayed put. On day 2, the family in apartment 1 was already in the right place. The remaining families from opposite ends exchanged apartments, namely, family 1 with family 6, family 2 with family 5, and family 3 with family 4.

Heracles was delighted that the big headache could be resolved in just two days.

Chapter 40

The Pandora Box

When the flood refugees finally went home, I helped Heracles clean out the mess they had left behind, as well as the pile of junk which had been removed from the Mount Olympus Warehouses to make room for the families.

I found a large cubical box with a button at each corner of the lid. The following was inscribed in the middle of the lid.

1. Each button is either on or off, but one cannot tell its state just by looking at it.

2. Whenever a button is pressed, its state changes.

3. Several buttons may be pressed simultaneously.

4. After each round of button-pressing, the box will spin rapidly about its vertical axis, so that afterwards one can no longer tell which button or buttons were just pressed.

5. The box will open when all buttons are in the same state.

"That is the famous Pandora Box, Eubie," exclaimed Heracles. "I have been looking for it for some time. I wonder how it ended up in the warehouses."

"I have never heard of it before, sir. Who is Pandora?"

"She is the wife of the titan Epimetheus, a grandson of Mother Earth. During Zeus's revolt, he and his middle brother Prometheus sided with the Olympians while his eldest brother Atlas led the titan army, as I had told you before. Prometheus was credited with creating us mortals out of river mud. Later, he entered the Mount Olympus Palace secretly and stole a glowing coal

181

from Hestia's hearth. He carried it safely down to earth so that we mortals could cook our meat instead of eating it raw."

"He is a benefactor of mankind," I said.

"Zeus wanted to cause trouble for us mortals that Prometheus had created. He sent Pandora to Epimetheus to be his wife. Prometheus, sensing a trick, warned his brother against accepting the bride. Zeus took Prometheus away for some cruel punishment. I won't go into the gruesome details. It was enough to frighten Epimetheus into obeying Zeus."

"What can you tell me about the Pandora Box?"

"Before he was taken away, Prometheus asked Epimetheus to keep a box safely hidden, and on no account to open it. Pandora found it one day, and with an instruction sheet given to her by Zeus, she opened it!"

"What happened?"

"She let loose a swarm of nasty winged things called Old Age, Sickness, Insanity, Spite, Passion, Vice, Plague, Famine and so forth. She herself was stung most viciously."

"Even worse," I observed, "these nasty winged things are still attack us and spoil everything for us. Perhaps I should get the box open, and get all of them back in?"

I tried to lift the lid off the box, but it did not move. That meant that the four buttons were not all in the same state.

"Pandora let go of the lid when she was stung," explained Heracles. "The instruction sheet fell inside the box, and then the lid slammed shut."

"I supposed what is on the sheet of opening instructions is the sequence in which to press the buttons to open the box," I said. "How can we open it without that sheet?"

"Hephaestus is the one we should consult. However, there is no urgency to get it open now. It will be a good challenge for you to see if you can figure it out on your own."

I tried for a while without making any progress. Meanwhile, Heracles still needed help. So I halted the mental work and did some physical work alongside him.

When I was ready to return to Eleusis, I asked Heracles for permission to take the Pandora Box with me. He did not object. So I took it back, and showed it to Silenus. He had just returned from his holidays, and had not finished unpacking as yet. So I worked at it on my own. Eventually, I figured out how to open the box. To my intense surprise, a bright-winged creature flew out.

"That is Hope!" exclaimed Silenus, who had just come over to see if he could help. "Apparently, she did not got out with the other creatures back then. Like Prometheus, Eubie, you have done mankind a tremendous service. Hope will keep mortals from killing themselves when they are in utter despair. How did you open the box?"

Solution

As usual, I simplified the problem by using fewer buttons. The case with only 1 button was trivial since the box would open automatically. If there were 2 buttons and the box did not open, pressing either button would do. Suppose there were 3 buttons. I realized that I only needed to press 1 button, because pressing 2 was exactly the same as just pressing the third one, and pressing all 3 was exactly the same as pressing no buttons at all.

If the box did not open initially, then the states of the buttons were divided 2 to 1. After the box had spun, I might press a button whose state was in the majority, thereby maintaining the 2 to 1 split forever. Thus I had no way to be sure that I could open the box.

This was also true if there were 5 buttons, or any odd number of buttons larger than 1. At first I believed that I could open the box if the number of buttons were even. Then I saw why the task was impossible with 6 buttons. If I ignored every other button, I still could not be sure that I could get the remaining 3 buttons to have the same state.

I now knew that if the number of buttons had an odd divisor greater than 1, then the box could not be opened with certainty. Only the powers of 2 remained to be considered.

Since the Pandora Box had 4 buttons, there were 3 operations I could perform.

(α) Press 2 opposite buttons.

(β) Press 2 adjacent buttons.

(γ) Press just 1 button.

I used a white circle to denote a button that was on, and a black circle to denote a button that was off. I divided the 16 possible combinations of states into four groups, as shown in the diagram below.

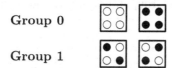

Group 0

Group 1

Group 2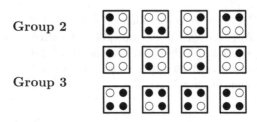

Group 3

Group 0 was called the *terminal* group since once I got there, I would not leave again because the box would have been opened. The combinations in Group 1 could be transformed into Group 0 by operation α. The combinations in Group 2 could be transformed into Group 0 or 1 by operation β. The combinations in Group 3 could be transformed into Group 0, 1 or 2.

My solution is presented in the diagram below. To fully appreciate the difficulty of the problem, it must be remembered that unless we are in the terminal group, there is no physical evidence for us to detect which group we may be in.

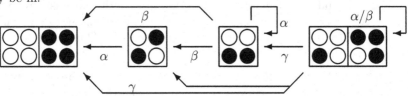

We try to open the box to see if we may be lucky enough to be already in the terminal group. If not, we will carry out the following 7 steps.

Step 1. Perform operation α.

If we are in Group 1, we would now be in Group 0 and the box will open. If we are in Group 2 or 3, we will remain there. In other words, if the box does not open after operation α, we have to be in Group 2 or 3.

Step 2. Perform operation β.

Suppose the box does not open. If we are in Group 2, we will now be in Group 1. If we are in Group 3, we will still be there. In any case, we will not be in Group 2.

Step 3. Perform operation α.

We must resist the temptation to proceed with Operation γ, because we may have gone back to Group 1. If the box does not open after operation α, then we will know that we are in Group 3.

Step 4. Perform operation γ.

This will take us from Group 3, and we will not return there.

Steps 5 to 7. Repeat Steps 1 to 3.

A succinct summary of the solution would be $\alpha\beta\alpha\gamma\alpha\beta\alpha$, the same as the solution I showed Eros in the "Tower of Delphi".

Appendix

Source of the Problems

In this appendix, we list the 40 problems from the International Mathematics Tournament of the Towns which are embedded into our mathematical puzzle tales. Each problem is identified by its season (Spring or Fall), its year, its paper (Junior or Senior), its level (Ordinary or Advanced) and its number in that competition. The problem number indicates the number of the chapter in which the problem can be found. In some cases, it has been modified to fit the story. The reader can see the effort taken in this task. They may also consult Robert Graves's book *Greek Gods and Heroes* and see the effort taken in modifying the story to fit the problem.

We will *not* present solutions to these 40 problems because most of them have been solved (as they are) in the main part of this book. In cases where the modifications have been significant, we leave something for the reader to chew on.

The solutions to problems from Spring 1980 to Spring 2007 may be found in the following six books published by the **Australian Mathematics Trust** under the leadership of **Peter J. Taylor**.

[1] Taylor, P. J. *International Mathematics Tournament of the Towns, Volume 1, 1980–1984*, AMT, Canberra, 1993.

[2] Taylor, P. J. *International Mathematics Tournament of the Towns, Volume 2, 1984–1989*, AMT, Canberra, 1992.

[3] Taylor, P. J. *International Mathematics Tournament of the Towns. Volume 3, 1989–1993*, AMT, Canberra, 1994.

[4] Storozhev, A. M. and Taylor, P. J. *International Mathematics Tournament of the Towns, Volume 4, 1993–1997*, AMT, Canberra, 2003.

[5] Storozhev, A. M. *International Mathematics Tournament of the Towns, Volume 5, 1997–2002*, AMT, Canberra, 2006.

[6] Liu, A. and Taylor, P. J. *International Mathematics Tournament of the Towns, Volume 6, 2002–2007*, AMT, Canberra, 2009.

The solutions to problems from Fall 2007 to Spring 2021 may be found in the book *Mathematical Recreations from the Tournament of the Towns* by A. Liu and P. J. Taylor, published in 2022 by **CRC Press/Taylor & Francis**.

Problem 1. (Fall 2001 Junior O-4/Senior O-3)
On an east-west shipping lane are ten ships sailing individually. The first five from the west are sailing eastwards while the other five ships are sailing westwards. They sail at the same constant speed at all times. Whenever two ships meets, each turns around and sails in the opposite direction. When all ships have returned to port, how many meetings of two ships have taken place?

Problem 2. (Spring 2020 Junior A-2)
A dragon has 41! heads and the knight has three swords. The gold sword cuts off one-half of the current number of heads of the dragon plus one more. The silver sword cuts off one-third of the current number of heads plus two more. The bronze sword cuts off one-fourth of the current number of heads plus three more. The knight can use any of the three swords in any order. However, if the current number of heads of the dragon is not a multiple of 2 or 3, the swords do not work and the dragon eats the knight. Will the knight be able to kill the dragon by cutting off all its heads?

Problem 3. (Spring 2017 Junior A-6)
On a $1 \times n$ board, a grasshopper can jump to the 8th, the 9th or the 10th square in either direction. Find an integer $n \geq 50$ such that starting from some square of the board, the grasshopper cannot visit every square exactly once.

Problem 4. (Spring 2016 Junior O-4/Senior O-3)
In a 10×10 board, the 25 squares in the upper left 5×5 sub-board are black, while all remaining squares are white. The board is dissected into a number of connected pieces of various shapes and sizes such that the number of white squares in each piece is three times the number of black squares in that piece. What is the maximum number of pieces?

Problem 5. (Fall 1985 Junior A-2)

(a) The game "Cat and Mouse" is played on a 4×4 board. The Cat starts at one of the four central squares and the Mouse at the other central square on the same diagonal. The Cat moves first, then the Mouse moves, and the moves alternate thereafter. In each move, the Cat or the Mouse can move between squares sharing a common side. The Cat catches the Mouse if it lands on the square where the Mouse is. Is there a strategy for the Cat to catch the Mouse?

(b) Answer the same question on the modified game board where a corner square is removed, but movement is allowed between the two squares sharing a common side with the corner square?

(c) Answer the same question if the same modification is made at the opposite corner as well.

Problem 6. (Fall 1997 Junior A-3)
Initially there is a checker on every square of a $1 \times n$ board. The first move consists of moving any checker to an adjacent square. This may be on either side if the checker is not at an end of the board. This creates a stack of two checkers. Each subsequent move consists of moving any stack in either direction, as many squares as the number of checkers in the stack, provided that it stays within the boundaries of the board. If a stack lands on a non-empty square, it is placed on top of the stack already there, forming a single stack. Prove that it is possible to stack all the checkers on one square in $n-1$ moves.

Problem 7. (Spring 2003 Junior A-5)
Some squares in a 9×9 board are to be cut along both diagonals. What is the largest number of squares that can be cut without the board falling apart into several pieces?

Problem 8. (Fall 1991 Junior A-5)
Some of the 81 squares in a 9×9 board are colored in such a way that the distance between the centers of any two colored squares is greater than 2.

(a) Does there exist such a coloring with 17 colored squares?

(b) Prove that there cannot be more than 17 colored squares.

Problem 9. (Spring 2014 Junior O-2)
Olga's mother baked 7 apple pies, 7 banana pies and 1 cherry pie. They are arranged in that exact order on a round plate when they are put into the microwave oven. All the pies look alike, but Olga knows only their relative positions on the plate because it has rotated. She wants to eat the cherry pie. She is allowed to taste three of them, one at a time, before making up her mind which one she will take. Can she guarantee that she can take the cherry pie?

Problem 10. (Fall 2008 Senior A-7)
A contest consists of 30 true or false questions. Victor knows nothing about the subject matter. He may write the contest several times, with exactly the same questions, and is told how many questions he has answered correctly each time. Can he be sure that he will answer all 30 questions correctly

(a) on his 30th attempt;

(b) on his 25th attempt?

Problem 11. (Spring 1984 Junior O-1/Senior O-1)
The price of 175 Humpties is more than the price of 125 Dumpties but less than that of 126 Dumpties. Prove that you cannot buy three Humpties and one Dumpty for

(a) 80 cents;

(b) 1 dollar.

Problem 12. (Fall 2013 Senior A-1)
Penny chooses an interior point of one of the 64 squares of a standard board. Basil draws a sub-board consisting of one or more squares such that its boundary is a single closed polygonal line which does not intersect itself. Penny will then tell Basil whether the chosen point is inside or outside this sub-board. What is the minimum number of times Basil has to do this in order to determine whether the chosen point is black or white?

Problem 13. (Fall 1998 Senior A-4)
The twelve members of a Jury were to be seated at a round table. Name plates were placed at each seat. Professor K., being absent-minded, took the the seat which is next to where he should be in the clockwise direction. One at a time, the other members occupied their correct seats, unless already occupied. In such a case, the member would take the first vacant seat in the clockwise direction. Of course, the final seating arrangement depends on the order in which the other members took their seats, after Prof. K.. How many different seating arrangements were possible?

Problem 14. (Spring 2014 Junior O-5/Senior O-3)
Forty Thieves are ranked from 1 to 40, and Ali Baba is also given the rank 1. They want to cross a river using a boat. Nobody may be in the boat alone, and no two people whose ranks differ by more than 1 may be in the boat at the same time. Is this task possible?

Problem 15. (Spring 1984 Senior A-3)
An infinite (in both directions) sequence of rooms is situated on one side of an infinite halfway. The rooms are numbered by consecutive integers and each contains a grand piano. A finite number of pianists live in these rooms. There may be more than one of them in some of the rooms. Every day some two pianists living in the k-th and $(k + 1)$-st rooms for some k decide that they interfere with each other's practice, and they move to the $(k − 1)$-st and $(k + 2)$-nd rooms, respectively. Prove that these moves will cease after a finite number of days.

Problem 16. (Spring 2021 Junior A-6/Senior A-5)
A hotel has n unoccupied rooms upstairs, k of which are under renovation. All doors are closed, and it is impossible to tell if a room is occupied or under renovation without opening the door. There are 100 tourists in the lobby downstairs. Each in turn goes upstairs to open the door of some room. If it is under renovation, she closes its door and opens the door of another room, continuing until she reaches a room not under renovation. She moves in that room and then closes the door. Each tourist chooses the doors she opens. For each k, determine the smallest n for which the tourists can agree on a strategy while in the lobby, so that no two of them move into the same room.

Problem 17. (Fall 1984 Junior A-4/Senior A-4)
Six musicians gathered at a chamber music festival. At each scheduled concert, some of these musicians played while the others listened as members of the audience. What is the minimum number of such concerts in order to enable each musician to listen, as a member of the audience, to all the other musicians?

Problem 18. (Fall 1995 Senior A-4)
Along a track for cross-country skiing, 1000 seats are placed in a row and numbered in order from 1 to 1000. By mistake, n tickets were sold, $100 < n < 1000$, each with one of the numbers $1, 2, \ldots, 100$ printed on it. Of course, many tickets will have the same seat numbers, and each of the numbers $1, 2, \ldots, 100$ appears on at least one ticket. The n spectators arrive one at a time. Each goes to the seat shown on his ticket and occupies it if still empty. If not, he just says "Oh" and moves to the seat with the next number. This is repeated until he finds an empty seat and occupies it, saying "Oh" once for each occupied seat passed over but not at any other time. Prove that the total number of times someone says "Oh" does not depend on the order in which the n spectators arrive, although it does depend on the distribution of the numbers on the tickets.

Problem 19. (The Fall 2019 Junior A-7)
Peter has an $n \times n$ stamp, $n > 10$, such that 102 of the squares are coated with black ink. He presses this stamp 100 times on a 101×101 board, each time leaving a black imprint on 102 unit squares of the board. Is it possible that the board is black except for one square at a corner?

Problem 20. (Spring 2008 Junior A-5)
Standing in a circle are 99 girls, each with a candy. In each move, each girl gives her candy to either neighbor. If a girl receives two candies in the same move, she eats one of them. What is the minimum number of moves after which only one candy remains?

Problem 21. (Spring 1981 Senior A-5)
The first quadrant is divided into a checkerboard infinite in two directions. On some squares there are pieces. It is possible to transform the positions of

the pieces according to the following rule: if the neighboring squares to the right and above a given piece are free, it is possible to remove this piece and put pieces on these free squares. Consider the bottom three squares in the first column, the bottom two in the second, and the bottom one in the third. The goal is to have these six squares free of pieces. Is it possible to reach this goal if in the initial position,

(a) there are 6 pieces and they are placed on the 6 "forbidden" squares;

(b) there is only one piece, located in the bottom square of the first column?

Problem 22. (Spring 1986 Junior A-6)
There are 1001 steps going up a hill, with one rock on each of the first 500 steps. Sisyphus may pick up any rock and moves it up to the next empty step. Then Aid may pick up any rock immediately above an empty step and moves it down to that empty step. Thereafter, Sisyphus and Aid continue their moves alternately. Sisyphus wants to place a rock on the top step. Can Aid stop him?

Problem 23. (Spring 1981 Junior A-4/Senior A-4)
Each of k friends simultaneously learns one different item of news. They begin to phone one another to tell them their news. Each conversation lasts exactly one hour, during which time it is possible for two friends to tell each other all of their news. What is the minimum number of hours needed in order for all of the friends to know all of the news, if

(a) $k = 64$;

(b) $k = 55$;

(c) $k = 100$.

Problem 24. (Fall 2009 Junior A-7/Senior A-6)
Olga and Max visited a certain Archipelago with 2009 islands. Some pairs of islands were connected by boats which run both ways. Olga chose the first island on which they land. Then Max chose the next island which they could visit. Thereafter, the two took turns choosing an accessible island which they had not yet visited. When they arrived at an island which was connected only to islands they had already visited, whoever's turn to choose next would be the loser. Prove that Olga could always win, regardless of how Max played and regardless of the way the islands were connected.

Problem 25. (Fall 2004 Junior O-2/Senior O-2)

(a) A bag contains 111 balls, each of which is green, red, white or blue. If 100 balls are drawn at random, there will always be 4 balls of different colors among them. What is the smallest number of balls that must be drawn, at random, in order to guarantee that there will be 3 balls of different colors among them?

(b) A bag contains 100 balls, each of which is red, white or blue. If 26 balls are drawn at random, there will always be 10 balls of the same color among them. What is the smallest number of balls that must be drawn, at random, in order to guarantee that there will be 30 balls of the same color among them?

Problem 26. (Spring 1990 Senior A-3)
A cake is prepared for a dinner party to which only p or q persons will come, p and q being relatively prime integers. Find the minimum number of pieces, not necessarily equal, into which the cake must be cut in advance so that the cake may be equally shared among the persons in either case.

Problem 27. (Fall 1997 Junior A-6/Senior A-6)
Points on each side of an equilateral triangle divide the side into n equal segments. Lines parallel to the sides of the triangle are drawn through all of these points, dividing the original triangle into n^2 small triangles or "cells". The cells between any two adjacent parallel lines form a "stripe". What is the maximum number of cells that can be chosen such that no two of them belong to a single stripe in any of the three orientations, if

(a) $n = 10$;

(b) $n = 9$?

Problem 28. (Fall 2017 Junior A-4/Senior A-1)
One hundred doors and one hundred keys are numbered 1 to 100 respectively. Each door is opened by a unique key whose number differs from the number of the door by at most one. Is it possible to match the keys with the doors in n attempts, where

(a) $n = 99$;

(b) $n = 75$;

(c) $n = 74$?

Problem 29. (Fall 2007 Junior A-6/Senior A-5)
The audience arranges n coins in a row. The sequence of heads and tails is chosen arbitrarily. The audience also chooses a number between 1 and n inclusive. Then the assistant turns one of the coins over, and the magician is brought in to examine the resulting sequence. By an agreement with the assistant beforehand, the magician tries to determine the number chosen by the audience.

(a) If this is possible for some n, is it also possible for $2n$?

(b) Prove that if this is possible for some n_1 and n_2, then it is also possible for $n_1 n_2$.

(c) Determine all n for which this is possible.

Problem 30. (Spring 2009 Junior A-5)
A castle is surrounded by a circular wall with 9 towers. Some knights stand on guard on these towers. After every hour, each knight moves to a adjacent tower. A knight always moves in the same direction, whether clockwise or counter-clockwise. At some hour, there are at least two knights on each tower. At another hour, there are exactly 5 towers each of which has exactly one knight on it. Prove that at some other hour, there is a tower with no knights on it.

Problem 31. (Fall 2019 Junior O-3/Senior O-5)
Counters numbered 1 to 100 are arranged in order in a row. It costs 1 dollar to interchange two adjacent counters, but nothing to interchange two counters with exactly k other counters between them. What is the minimum cost for rearranging the 100 counters in reverse order, if

(a) $k = 3$;

(b) $k = 4$?

Problem 32. (Spring 2008 Senior A-6)
Seated in a circle are 11 wizards. A different positive integer not exceeding 1000 is pasted onto the forehead of each. A wizard can see the numbers of the other 10, but not his own. Simultaneously, each wizard puts up either his left hand or his right hand. Then each declares the number on his forehead at the same time. Is there a strategy on which the wizards can agree beforehand, which allows each of them to make the correct declaration?

Problem 33. (Spring 2011 Senior A-7)
Among a group of programmers, every two either know each other or do not know each other. Eleven of them are geniuses. Two companies hire them one at a time, alternately, and may not hire someone already hired by the other company. There are no conditions on which programmer a company may hire in the first round. Thereafter, a company may only hire a programmer who knows another programmer already hired by that company. Is it possible for the company which hires second to hire ten of the geniuses, no matter what the hiring strategy of the other company may be?

Problem 34. (Fall 1991 Junior A-7/Senior A-7)
There are n children who want to share equally m identical pieces of chocolate, without sharing any piece between more than two children.

(a) If $m = 9$, for what values of n is this possible?

(b) For what values of m and n is this possible?

Problem 35. (Spring 1991 Junior A-6/Senior A-6)
In a tournament, each boxer can fight no more often than once per day. It is known that the boxers are of different strengths, and the stronger boxer

always wins. The schedule of fights for each day is determined on the evening before, and cannot be changed during the day. Prove that the boxers can be sorted out in order of strength in

(a) 10 days if there are 16 boxers;

(b) 15 days if there are 32 boxers.

Problem 36. (Spring 2001 Senior A-2)
For any group of at least 5 students, 80% of the F grades received by the students in this group are given to at most 20% of them. Prove that at least 75% of the F grades received by all the students are given to just one of them.

Problem 37. (Fall 2005 Senior A-6)
A blackboard is initially empty. In each move, one may either add two 1s, or erase two copies of a number n and replace them with $n-1$ and $n+1$. What is the minimum number of moves needed to put 2005 on the blackboard?

Problem 38. (Spring 1981 Junior A-5/Senior A-2)
A game is played on an infinite plane. There are two players. One has a piece known as a "wolf", while the other has k pieces known as "sheep". The first player moves the wolf, then the second player moves a sheep, the first player moves the wolf again, the second player moves a sheep, and so on. The wolf and the sheep can move in any direction, with a maximum distance of one meter per move. Is it true that there exists an initial position from which the wolf cannot capture any sheep if

(a) $k = 50$;

(b) k is any positive integer?

Problem 39. (Spring 1987 Senior A-3)
In a certain city, only simple pairwise exchanges of apartments are allowed. If two families exchange apartments, they are not allowed to participate in another exchange on the same day. It is assumed that a family does not split up, and that exactly one family occupies each apartment before and after each exchange. Prove that any compound exchange may be effected in two days.

Problem 40. (Fall 2009 Senior A-7)
At the entrance to a cave is a rotating round table. On top of the table are n identical barrels, evenly spaced along its circumference. Inside each barrel is a herring either with its head up or its head down. In a move, Ali Baba chooses from 1 to n of the barrels and turns them upside down. Then the table spins around. When it stops, it is impossible to tell which barrels have been turned over. The cave will open if the heads of the herrings in all n barrels are all up or are all down. Determine all values of n for which Ali Baba can open the cave in a finite number of moves.

Answers

We reemphasize that we will not be presenting *solutions* here, only answers. These will be given in *chronological* order rather than in *numerical* order. This is done deliberately to diminish the chances that the reader may catch sight of the answer to the next problem which is about to be attempted. A few of the problems ask for proofs and there will not be any answers.

Season/Year	Problem	Answers
Spring 1981	21	(a) No; (b) No.
	23	(a) 6; (b) 7; (c) 7.
	38	(a) Yes. (b) Yes.
Fall 1984	17	4.
Fall 1985	5	(a) No; (b) Yes; (c) No.
Spring 1986	22	Yes.
Spring 1990	26	$p + q - 1$.
Fall 1991	8	(a) Yes.
	34	(a) 1,2,3,4,5,6,7,8,9,10,12,18;
		(b) $n \le m$ or $n = m + k$, k divides m.
Fall 1997	27	(a) 7; (b) 6.
Fall 1998	13	1024.
Fall 2001	1	25.
Spring 2003	7	21.
Fall 2004	25	(a) 88; (b) 66.
Fall 2005	37	1342355520.
Fall 2007	29	(c) All powers of 2.
Spring 2008	20	98.
	32	Yes.
Fall 2008	10	(a) Yes; (b) Yes.
Fall 2009	40	All powers of 2.
Spring 2011	33	Yes.
Fall 2013	12	2.
Spring 2014	9	Yes.
	14	Yes.
Spring 2016	4	9.
Spring 2017	3	63.
Fall 2017	28	(a) Yes; (b) Yes; (c) No.
Fall 2019	19	Yes.
	31	(a) 50; (b) 61.
Spring 2020	2	Yes.
Spring 2021	16	$50k + 100$ if $k = 2t$ for some $t \ge 1$;
		$50k + 51$ if $k = 2t + 1$ for some $t \ge 0$.